*In memory of **ISAAC YAGLOM**,
the great expositor of
the world of mathematics.*

INTRODUCTIONS

This interesting and delightful book by two well-known geometers is written both for mature mathematicians interested in somewhat unconventional geometric problems and especially for talented young students who are interested in working on unsolved problems which can be easily understood by beginners and whose solutions perhaps will not require a great deal of knowledge but may require a great deal of ingenuity.

Many unsolved problems are discussed; for example, about tiling of squares with polyominoes and also many exercises are given of various degrees of difficulty.

There is also an interesting chapter on existence proofs, the understanding of which perhaps requires more mathematical maturity. There is also a chapter on graph theory and a slightly more difficult chapter on the Jordan Theorem.

There is also a more difficult chapter on combinatorial geometry where the famous unsolved conjecture of Borsuk is discussed in great detail. Fifty years ago I spent lots of time trying to prove it. To quote Hardy, I hope younger and stronger hands (or rather brains) will have more success.

The last two chapters deal with illumination problems and Helly and Szökefalvi-Nagy's theorem. Here also, many unsolved problems are stated.

I recommend this book very warmly.

PAUL ERDÖS
Member of the Hungarian
Academy of Sciences
Honorary Member of the
National Academy of
Sciences of the USA

January, 1991
Gainesville, Florida

VLADIMIR BOLTYANSKI

*National Research Institute
of System Research
USSR*

and

ALEXANDER SOIFER

*University of Colorado
at Colorado Springs
USA*

GEOMETRIC ETUDES
IN
COMBINATORIAL
MATHEMATICS

With over 300 Illustrations, Index, and Introductions by

*PAUL ERDÖS
BRANKO GRÜNBAUM
and CECIL ROUSSEAU*

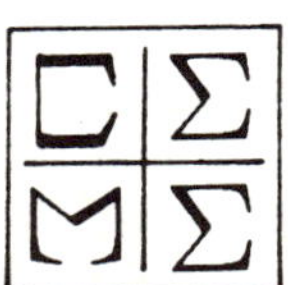

**Center for Excellence
in Mathematical Education
Colorado Springs, 1991**

Center for Excellence in Mathematical Education
885 Red Mesa Drive
Colorado Springs, Colorado 80906, USA

Authors: Vladimir Boltyanski
National Research Institute of System Research
9 pr. 60-letia Oktyabrya
117312 Moscow, USSR

Alexander Soifer, Professor of Mathematics
College of Letters, Arts and Sciences
University of Colorado
Colorado Springs, Colorado 80933, USA

Cover: Yuri Soifer

Illustrations: Alexander Soifer

First Edition

Printed in the United States of America

AMS Subject Classification (1980):
52-01, 52-02, 52A10, 52A15, 52A35, 52A45, 05C99, 00A07
Library of Congress Catalog Card Number 90-086323
ISBN 0-940263-02-5

How do young people develop skills of any kind--from driving cars to playing basketball or a musical instrument? In all cases the sequence of events is the same: a little instruction, more or less formal, is followed by ample practice. The person wishing to acquire better skills must invest, on his or her own, considerable efforts aimed at gaining better mastery of various aspects of the activity.

Mathematics in general, and geometry in particular, are also fields in which the small amount of formal instruction (often very superficial) given in schools is not sufficient to bring latent talents to full development. Individual work and effort are necessary, and one of the shortcomings of our educational system is the lack of extracurricular material that would make such independent study of geometry (or other mathematics) attractive and interesting.

The present book is an appealing step in the direction of providing useful supplementary reading and practice material. It is intended for the use of our high school students--it was written with that audience in mind, and is aimed at giving accessible but not trivial opportunities for exercising the geometric intuition as well as deductive reasoning. The discussion is self-contained and easy to follow--but despite the elementary character of the mathematics involved, there are plenty of challenging questions, and even several open problems that are easy to state but have so far resisted all attempts to solve them.

The authors build on the tradition and experience of the educational system of the U.S.S.R., in which books of this nature have long played an important role. This represents popularization of science at its best. Many contemporary mathematicians (the writer of these lines included, and still appreciative of the experience) obtained their first taste of the geometry of convex figures from a book very similar in spirit to the present one, which was coauthored by Professor Boltyanski, one of the authors of this work. (It is listed under [YB] in the Bibliography).

Throughout the text, the authors show great mastery of the topics discussed. Their infectious enthusiasm for opening our eyes to the beauties of the worlds of geometry and combinatorics should make this book attractive to wide audience. It is to be hoped that the following pages will bring the joy of understanding, seeing and discovering geometry to many of our young people. It is also to be hoped that many other texts of a similar nature will follow, to help lift our teaching out of the present doldrums.

BRANKO GRÜNBAUM
Professor of Mathematics
University of Washington

February, 1991
Seattle, Washington

Some areas of mathematics are not as well known as they deserve to be. They are like the out-of-the-way places where a discerning traveler can find unexpected pleasure and satisfaction. Combinatorial geometry is such a mathematical area. Its basic ideas are easily within the grasp of a bright high school student. However, there are many trained mathematicians who are unaware of even its most essential problems and achievements.

GEOMETRIC ETUDES IN COMBINATORIAL MATHE-MATICS provides the reader an opportunity to explore this beautiful area of mathematics. Expertly guided by Alexander Soifer and Vladimir Boltyanski, the reader is surprised and delighted by exquisite gems of geometry and combinatorics. A leisurely and captivating presentation leads the reader into a world of tilings, graphs, and convex figures. It is a world that will be long remembered for its striking problems and results.

CECIL ROUSSEAU
Professor of Mathematics
Memphis State University
Coach of American Team
for the International
Mathematics Olympiad

January, 1991
Memphis, Tennessee

CONTENTS

INTRODUCTIONS iv

TABLE OF CONTENTS viii

PREFACE x

CHAPTER I **Tiling a Checker Rectangle** 1
    1. Introduction 1
    2. Tiling Rectangles by Trominoes 5
    3. Tetrominoes and "Color" Reasoning 14
    4. Tiling by Linear Polyominoes 25
    5. Polyominoes and Rotational
       Symmetries 37
    6. Tiling on Other Surfaces 50

CHAPTER II **Proofs of Existence** 65
    7. The Pigeonhole Principle in Geometry . . . 65
    8. An Infinite Flock of Pigeons 71

CHAPTER III **A Word About Graphs** 90
    9. Combinatorics of Acquaintance, or
       Introduction to Graph Theory 90
   10. More About Graphs 100
   11. Planarity 116
   12. The Intersection Index and the
       Jordan Theorem 125

CHAPTER IV **Ideas of Combinatorial Geometry** 136
   13. What are Convex Figures? 136
   14. Decomposition of Figures Into Parts
       of Smaller Diameters 155
   15. Figures of Constant Width 164

Chapter IV   (Continued)

    16.   Solution of the Borsuk Problem
          for Figures in the Plane 180

    17.   Illumination of Convex Figures 191

    18.   Theorems of Helly and Szökefalvi-Nagy .. 207

BIBLIOGRAPHY 226

INDEX 230

NOTATIONS 235

PREFACE

> *"The soul of every mathematician is wrestled for by the Devil of Abstract Algebra and the Angel of Topology."*
> *--Hermann Weyl*

Mathematics is frequently divided into elementary and higher mathematics, just like literature is divided into children's and grown-up's literature. We do not quite agree with this "discrimination" based on age. It would be more productive if we were to divide both mathematics and literature into good and not so good. Accordingly, we decided to make some grown-up's mathematics available to young mathematicians and their teachers.

Joy of creation, depth and beauty of ideas, flight of fantasy, and unexpected elegance of reasoning, which are so characteristic of mathematics, often remain outside of textbooks. Only popular books and mathematical olympiads enable students to peek into the Wonderland of Mathematics.

The intellectual eye of a child opens there to new and unusual problems, shining summits of magnificent new theories, unexpected "bridges" connecting these summits with each other and uniting them in one Wonderland of Mathematics. And most importantly, there is the joy of creating the opportunity to discover new, unexplored corners in the world of mathematics and then to notice with surprise that there is an unexpected path from these behind-the-cloud peaks--that opened up the intellectual eye--to the real world of things and happenings, to creating new machines and instruments, to solutions of life's problems--problems that previously seemed hopeless.

The main problem of popular literature is in opening, for an interested student, the mysterious world of contemporary mathematics, and, moreover, in bringing him to the forefront of this peaceful battle where he will be able join with prominent scientists in the fight for bringing out of the unknown new facts, ideas, and methods. Let these discoveries at first be small, but let them be. The only way for that is to work, to solve problems, and to overcome difficulties. We offer to you, our reader, not easy reading entertainment, but work and activity that calls forward and inspires.

We both, the authors of this book, love geometry. It is a remarkable region in the Wonderland of Mathematics. Moreover, geometry is not only an important part of the science; for us (as for the majority of mathematicians) geometry is a unique perception of the world that shines a bright light on unexplored ways in other areas of mathematics.

It often happens that while solving problems from algebra, analysis, logic or combinatorics, a mathematician draws in front of his intellectual eye a geometric picture that becomes more and more clear, detailed, and understandable--and suddenly geometric insight clears up completely an algebraic or combinatorial problem. The mathematician sits down at the table and writes dozens of formulas, integrals, and equations leading to the goal, to the solution of a new problem, a problem that is not at all geometric in its context. Without geometric ideas and representations, the mathematician would have long and painfully searched for a solution, like a blind kitten, losing the road, getting into dead ends, or senselessly wandering in circles.

We would be very happy if this book gives the reader the opportunity to broaden a little his geometric horizon, to get to believe in the magical strength of geometric ideas in the unending world of mathematics. As for combinatorics, probably no mathematician today can formulate precisely what combinatorics is and what problems and methods should be considered combinatorial. But more and more mathematicians invest their efforts in the development of new combinatorial directions in mathematics.

We invite young mathematicians to join this movement, this journey for discovering the "New World." The joy of creating, stubborn work, and the ability to cheer up if everything does not come out right from the beginning are your main tools in this journey in which there are no losers, but there are only winners.

And now a few words about us and how this book was created.

Who are we? A Soviet and an American mathematician. We got together in beautiful Colorado Springs and in a few concentrated weeks of writing, discussing, and problem solving long hours of each day and night, we produced the first rough draft of this book. It then took the second author eight months of editing, proofing, and adding some new material to bring this book to its final form.

This book discusses a few areas of combinatorial mathematics that have something in common. That something is a geometric flavor that we believe adds a visual appeal and distinctive beauty to mathematical reasoning. All four chapters: Tiling, Proofs of Existence, Graphs, and Combinatorial Geometry show that there is no border between problems of mathematical olympiads and research problems of mathematics. They introduce our young reader to some exciting ideas and concepts that are not easily available to them from other sources.

We hope life will enable us to continue our joint efforts in the future. We hope to produce a whole library of books for young and talented mathematicians.

We are grateful to Philip Engel, Paul Erdös, Martin Gardner, Branko Grünbaum, and Cecil Rousseau for being the first readers of our manuscript and providing us with valuable feedback. We are honored that Paul Erdös, Branko Grünbaum, and Cecil Rousseau have written introductions for this book.

Our friend and secretary, Lynn Scott, had to put up with two handwritings, one written all over the other (that is what comes out of joint efforts!). Thank you, Lynn!

Lilia Pashkova-Boltyanski took good care of our diet as we worked long hours on the book. Maya Soifer provided valuable help in producing illustrations. Thank you, wives!

We want a dialogue with you, our reader. Beautiful solutions, new problems, or whatever comes from your reading of our book interests us a great deal. Please share it all with us!

VLADIMIR BOLTYANSKI
National Research Institute
  of System Research
9 pr. 60-letia Oktyabrya
117312 Moscow
Russia

ALEXANDER SOIFER
University of Colorado
P.O. Box 7150
Colorado Springs, CO 80933
United States of America

January, 1991

CHAPTER I

TILING A CHECKER RECTANGLE

1. INTRODUCTION

Imagine you have a $m \times n$ rectangle R and lots of dominoes (a *domino* is a 1×2 rectangle). It is easy to find the conditions under which R can be *tiled* by dominoes, i.e., covered by dominoes, without any dominoes overlapping or sticking outside of the boundary of R. Indeed, R can be tiled by dominoes if and only if mn is even (prove it!).

The problem becomes a bit more difficult if we want to tile the same rectangle with exactly two monominoes (a *monomino* is a 1×1 square) and lots of dominoes (Figure 1.1) Where can these two monominoes be placed?

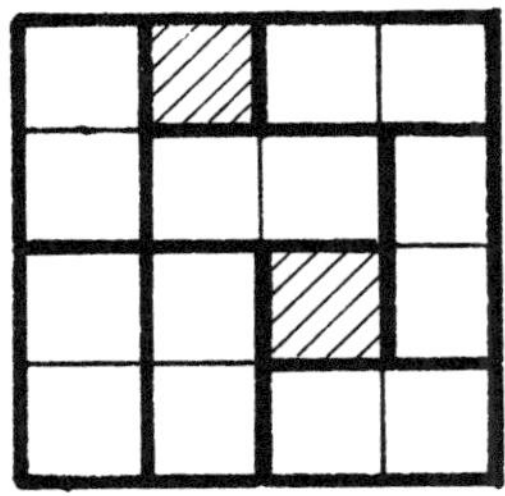

Figure 1.1

In order to answer this question we color the rectangle R in a chessboard fashion in two colors (Figure 1.2).

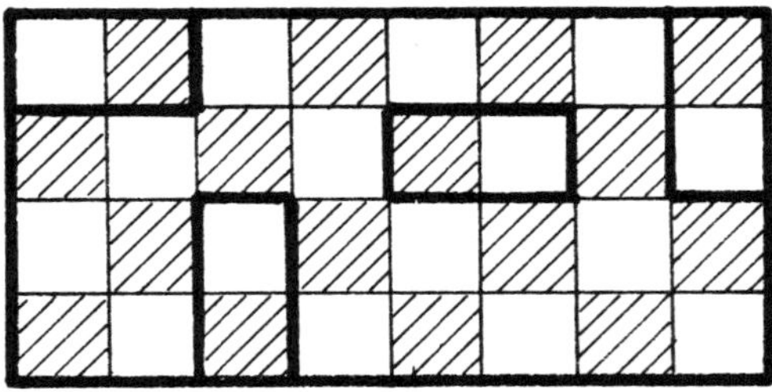

Figure 1.2

This coloring has a very nice property: regardless of how a domino is placed on the board, horizontally or vertically, it will cover exactly one square of each color (Figure 1.2). Therefore, the two monominoes *must* cover squares of different colors.

Gomery found a beautiful way to prove that conversely, no matter where the two monominoes are placed on the $m \times n$ board (where mn is even and both m and n are greater than 1) as long as they are on different colors the rest of the board can be tiled by dominoes.

Here is his proof: supposing n to be even (note at least one of the numbers m,n must be even), he created a labyrinth out of the board (Figure 1.3).

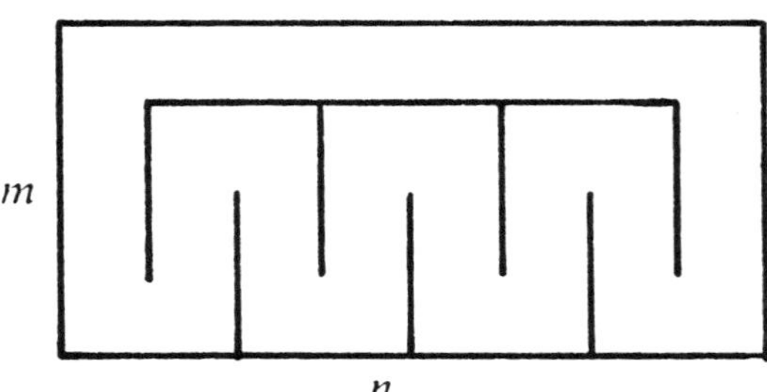

Figure 1.3

As you walk through this labyrinth, black and white squares alternate. Now let us cut the board along the walls of the labyrinth: we'll get a checkered ring with black and white squares alternating (Figure 1.4).

Figure 1.4

It is clear that no matter what single black and single white squares we cover by monominoes, the numbers of squares between them must be even and therefore the rest of the ring can be tiled by dominoes.

Solomon Golomb in his pioneering 1954 paper ([Go1]*) generalized the notion of domino (see also his book [Go2]). He named *n-omino* a figure made up of n squares of a checkerboard that are rook-connected, i.e., you can move from any square of the n-omino to any other by horizontal and vertical moves within the figure itself.

You are already familiar with the monomino and domino (Figure 1.5). In Figure 1.6 you can find all shapes of *trominoes* (3-minoes).

Exercise 1.1. Find all shapes of tetrominoes (4-minoes).

Exercise 1.2. Find all shapes of pentominoes (5-minoes).

---

*Here and everywhere in this book, symbols in brackets like [Go1] are references to the Bibliography found at the end of this book.

4

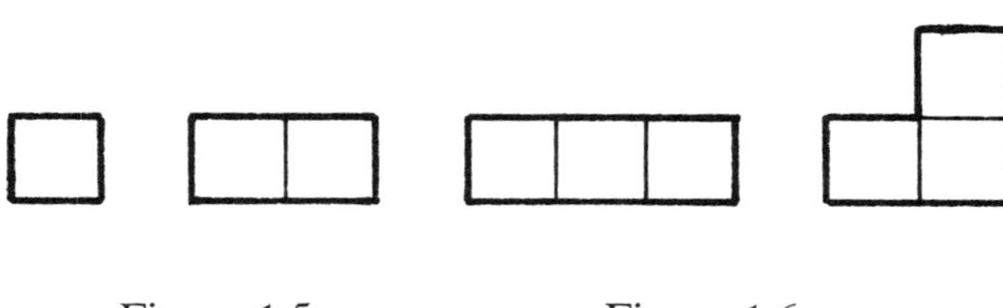

Figure 1.5   Figure 1.6

Solutions of Exercises

1.1. See Figure 3.1 below.

1.2. The answer is given in Figure 1.7 (which is taken from Golomb's book [Go2]. In this figure all twelve shapes of pentominoes are used to tile the 8×8 square with a 2×2 hole in the middle.

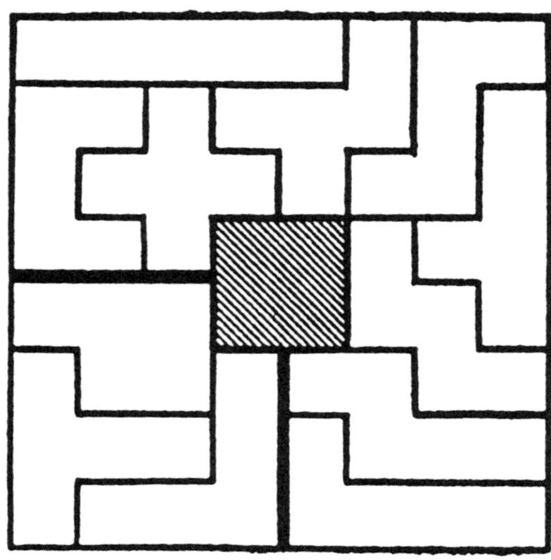

Figure 1.7

2. TILING RECTANGLES BY TROMINOES

Tiling a figure by tiles of an indicated shape (or, what is the same, cutting of a figure into parts of a given form) is a very interesting area of combinatorial geometry. It includes many fascinating exercises and research problems. Some of them we will consider in this section.

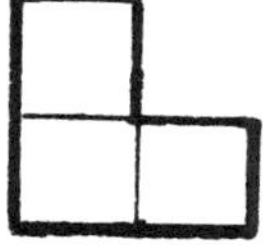

Figure 2.1

Here is a **L-tromino** (Figure 2.1). It is composed of three unit squares. (Note that there is one other tromino, the **linear tromino**, connecting three squares in a row. We will consider this tromino in Section 4.) We will consider in this section the following problem: *How to tile a rectangle by L-trominoes.* The first question is how to tile the 2×3 rectangle by several *L*-trominoes. A solution is trivial (Figure 2.2).

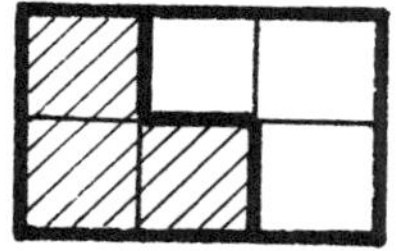

Figure 2.2

We are now going to offer you a sequence of exercises. You are invited to find solutions. After several attempts (successful or not) we recommend you either look through, or read attentively, our solutions at the end of the section.

6

Exercise 2.1. Is it possible to tile the 4×5 rectangle by L-trominoes?

Exercise 2.2. Find the smallest square that can be tiled by L-trominoes.

Exercise 2.3. Find all integers b such that the $2 \times b$ rectangle can be tiled by L-trominoes.

Exercise 2.4. Find all integers b such that the $3 \times b$ rectangle can be tiled by L-trominoes?

Combining the solutions of Exercises 2.3 and 2.4, we obtain the following result.

Theorem 2.1. *Let a,b be integers, such that $2 \leq a \leq 3$ and $b \geq a$. The $a \times b$ rectangle can be tiled by L-trominoes if and only if ab is divisible by 6.*

Equivalently: *The $a \times b$ rectangle with $2 \leq a \leq 3$ and $b \geq a$ can be tiled by L-trominoes if and only if one of the numbers a,b is divisible by 2 and the other is divisible by 3.*

Now a further direction of our research is clear: we will consider the $a \times b$ rectangles *in cases when $4 \leq a \leq b$.* But first we would like to share with you a couple of "philosophical" observations.

Suppose we are trying to solve a combinatorial problem. For example, we are trying to decide whether it is possible to tile a given figure F by L-trominoes. Let us assume further that we expect to receive the answer "no". How to get it? If some attempts to tile F with L-trominoes were unsuccessful, then we may not yet give the answer "no". Indeed, who knows, maybe the next attempt would be more successful. So, how to prove the answer "no"? In other words, how to establish that F cannot be tiled by L-trominoes? There are, in general, two ways. The first way consists of using an **invariant** property of considered figures. More precisely, the area of the L-tromino is equal to 3. So, each figure that can be tiled by L-trominoes has the area divisible by 3. This is an invariant, general property of all figures tiled by L-trominoes. Hence, if the figure F does not possess this property (that is, if its area is not divisible by 3), then F cannot be tiled by

L-trominoes. Such an approach is used in the solution of Exercise 2.1 (and some other exercises). The second way of getting the answer "no" consists of looking at all possibilities. Such an approach is used in the solution of Exercise 2.2 (see Figure 2.4). These two approaches of getting the answer "no" are most common.

Let us now suppose that we expect to receive the answer "yes" (to the question, whether the figure F can be tiled by L-trominoes). In order to get this answer, it is necessary to indicate a concrete tiling of the figure F by L-trominoes (or an algorithm for obtaining such a tiling).

You may be wondering what to do if we do not know what answer to expect. In this case we need to aid our intuition by *experimenting*, i.e., by considering and solving particular, relatively small and simple, cases of the problem.

At this point you, our reader, are well armed to prove Theorem 2.2. We created the following sequence of exercises to make your climbing to the summit of Theorem 2.2 easier.

Exercise 2.5. Prove that the 5×6 rectangle can be tiled by L-trominoes.

Exercise 2.6. Prove that the 5×9 rectangle can be tiled by L-trominoes.

Exercise 2.7. Prove that the 9×9 square can be tiled by L-trominoes.

Exercise 2.8. Prove that if $b > 5$ and b is divisible by 3, then the $5 \times b$ rectangle can be tiled by L-trominoes.

Exercise 2.9. Prove that if b is divisible by 3 and the $a \times b$ rectangle can be tiled by L-trominoes, then the rectangle $(a + 2) \times b$ can be tiled by L-trominoes.

Exercise 2.10. Integers a,b satisfy the conditions $a \geq 4$, $b \geq 5$, b is divisible by 3. Decide whether the $a \times b$ rectangle can be tiled by L-trominoes.

With the help of the solution of Exercise 2.10 we can obtain the following result:

Theorem 2.2. *Let F be the a × b rectangle where a $\geq$ 4, b $\geq$ 4. Then F can be tiled by L-trominoes if and only if the product ab is divisible by 3.*

Finally, we can give a combined statement of Theorems 2.1 and 2.2. *Let a,b be integers such that 2 $\leq$ a $\leq$ b. The a × b rectangle can be tiled by L-trominoes in and only in the following cases: i) a = 3 and b is even; ii) a $\neq$ 3 and ab is divisible by 3.*

Solutions of Exercises

2.1. The area of the *L*-tromino (see Figure 1.8) is equal to 3. The area of the 4 × 5 rectangle is equal to 20. Hence, in order to tile this rectangle by *L*-trominoes, it is necessary to use $\frac{20}{3}$ *L*-trominoes, which is impossible.

Generalizing, we obtain that if the $a \times b$ rectangle can be tiled by *L*-trominoes, then *ab* is divisible by 3. It is a necessary condition of such a tiling.

2.2. According to the solution of Exercise 2.1, if the $a \times a$ square can be tiled by *L*-trominoes, then *a* is divisible by 3. The 6 × 6 square can be decomposed into 2 × 3 blocks (Figure 2.3), and consequently, the 6 × 6 square can be tiled by *L*-trominoes (see Figure 2.2). So, it remains to find out whether the 3 × 3 square can be tiled by *L*-trominoes. The answer is "no". Indeed, all the possibilities to cover the left upper corner are shown in Figure 2.4.

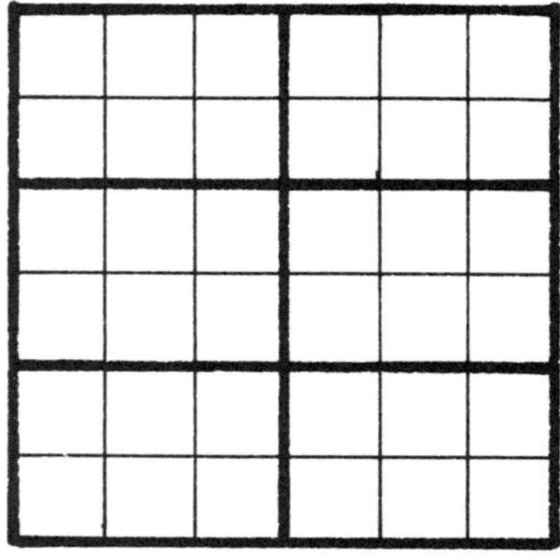

Figure 2.3

In the case of Figure 2.4a it is impossible to cover the upper right corner. In the case of Figure 2.4b it is impossible to cover the lower left corner. Finally, in the case of Figure 2.4c the only possibility to cover the right upper corner is shown in Figure 2.5. But then the three bottom squares remain uncovered. Thus the investigation of all possibilities shows that the 3×3 square cannot be tiled by L-trominoes. Consequently, the 6×6 square is the smallest tileable square.

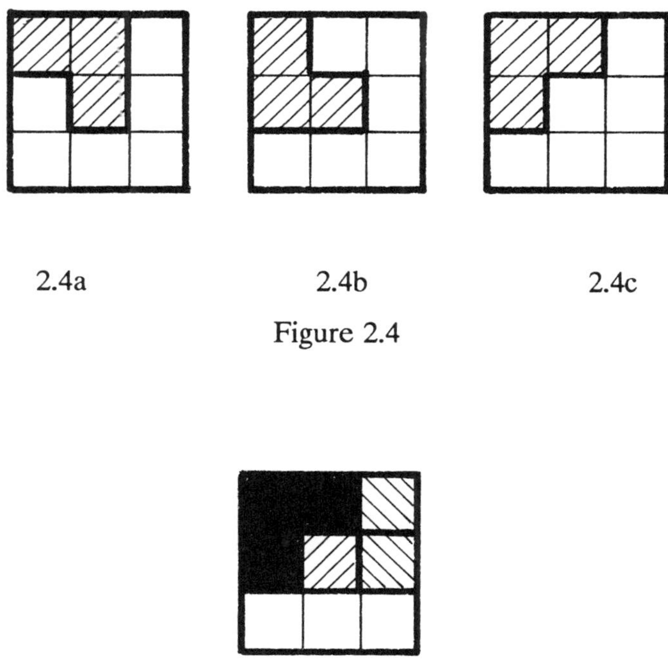

2.4a   2.4b   2.4c

Figure 2.4

Figure 2.5

2.3. If b is divisible by 3, then the $2 \times b$ rectangle can be decomposed into 2×3 blocks (see Figure 2.6) and, consequently, this rectangle can be tiled by L-trominoes. The reasoning in the solution of Exercise 2.1 shows that there are no other possibilities. So, the $2 \times b$ rectangle can be tiled by L-trominoes if and only if b is divisible by 3.

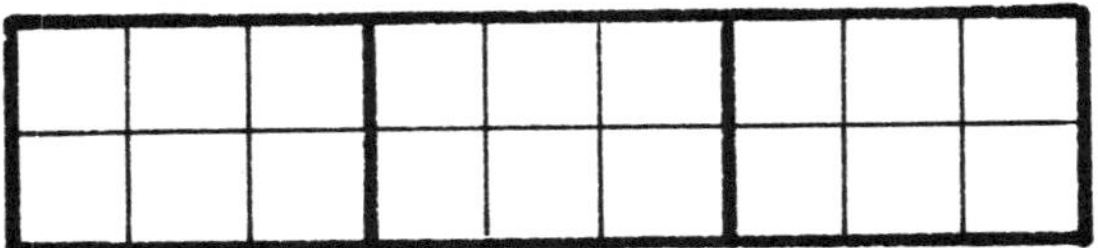

Figure 2.6

2.4. Let us consider a $3 \times b$ rectangle. All the possibilities to cover the left upper corner are shown in Figure 2.7. But in Figure 2.7c it is impossible to cover the lower left corner. As to possibilities in Figures 2.7a and 2.7b, the lower left corner can be covered uniquely. This is shown in Figure 2.8

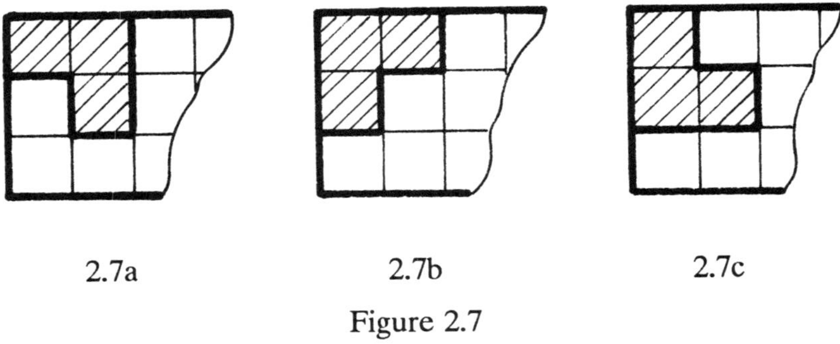

2.7a   2.7b   2.7c

Figure 2.7

Figure 2.8

In either case, we have the 3×2 block that is situated at the left side of the $3 \times b$ rectangle covered by two L-trominoes. In the same manner we obtain the next 3×2 block covered by two L-trominoes (Figure 2.9) and so on.

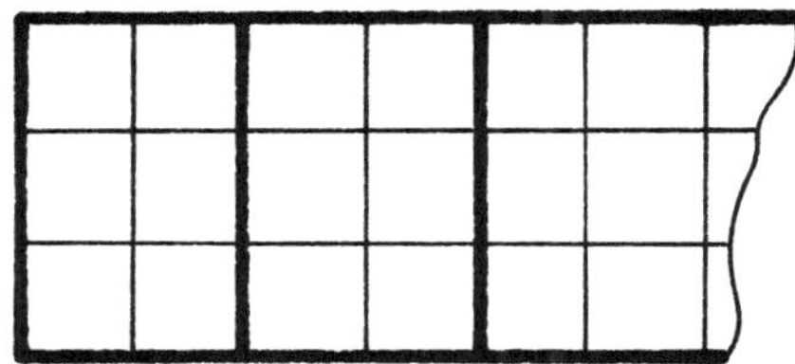

Figure 2.9

This reasoning shows that if the $3 \times b$ rectangle can be tiled by L-trominoes, then b is even. This condition is not only necessary, but also sufficient: if b is even, then the $3 \times b$ rectangle can be decomposed into 3×2 blocks.

2.5. The 5×6 rectangle can be decomposed into 2×3 blocks (Figure 2.10). Thus the rectangle can be tiled by L-trominoes.

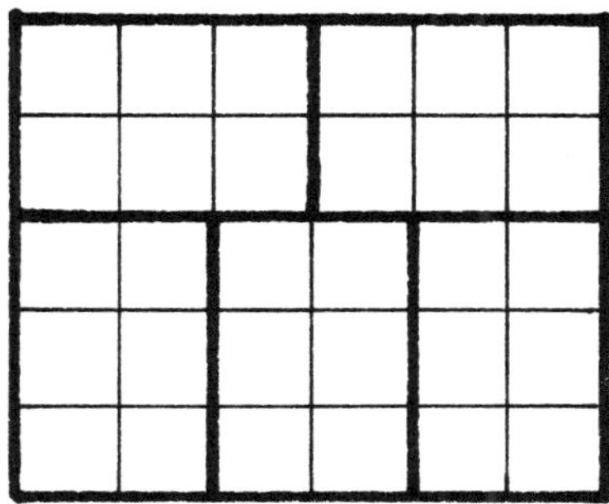

Figure 2.10

2.6. A solution is shown in Figure 2.11. Certainly this decomposition of the 5×9 rectangle into L-trominoes is not unique. Please note that in this exercise we have the $a \times b$ rectangle with $a = 5$, $b = 9$, so *ab is not* divisible by 6 (only by 3), and yet the rectangle is tileable by L-trominoes.

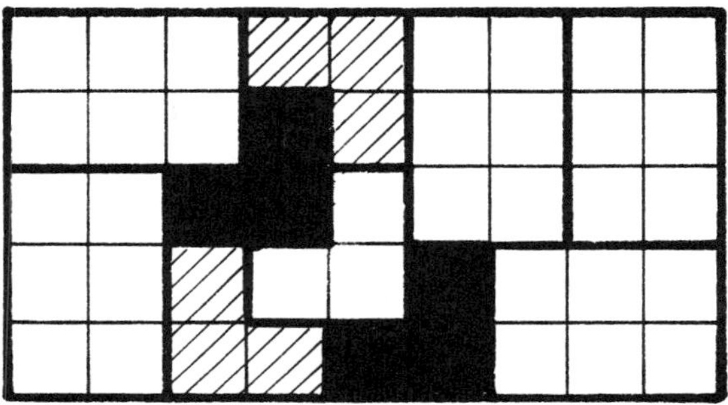

Figure 2.11

2.7. The 9×9 square can be decomposed into two rectangles 4×9 and 5×9 (Figure 2.12). Each of these rectangles can be tiled by L-trominoes (according to the results of Exercises 2.3 and 2.6). Thus the 9×9 square can be tiled by L-trominoes.

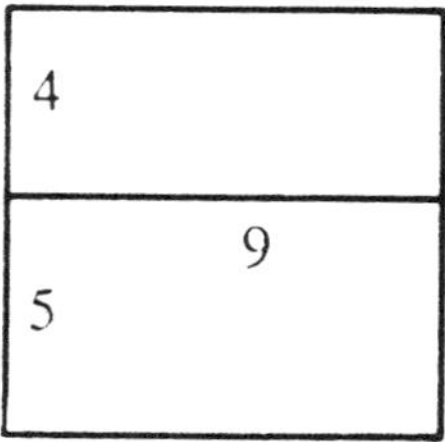

Figure 2.12

2.8. Each $5 \times b$ rectangle where b is divisible by 3 and $b \geq 6$ can be decomposed into blocks 5×6 and 5×9. Indeed, if b is divisible by 6, then it suffices to use only 5×6 blocks (Figure 2.13). And if b is not divisible by 6, then it is necessary to use one 5×9 block (see Figure 2.14).

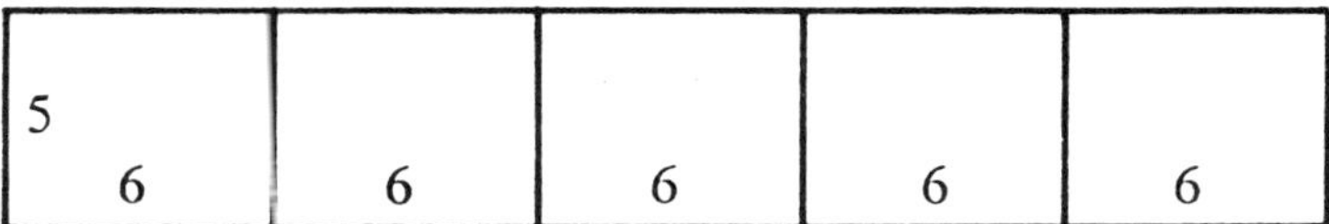

Figure 2.13

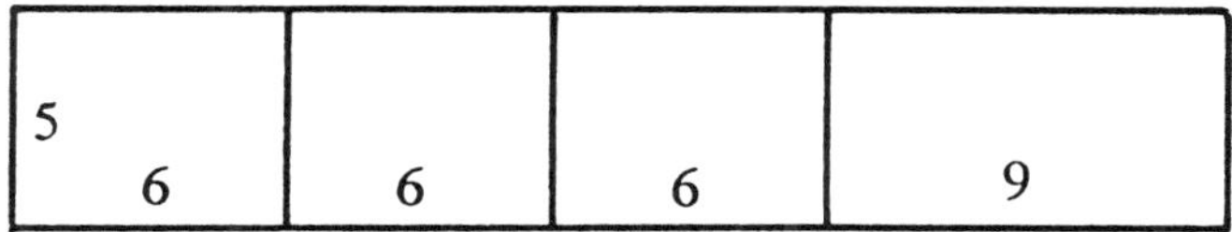

Figure 2.14

2.9. A solution is sketched in Figure 2.15. Since b is divisible by 3, then the $2 \times b$ rectangle can be decomposed into 2×3 blocks. So the $2 \times b$ rectangle can be tiled by L-trominoes. Therefore the $(a + 2) \times b$ rectangle can also be tiled by L-trominoes.

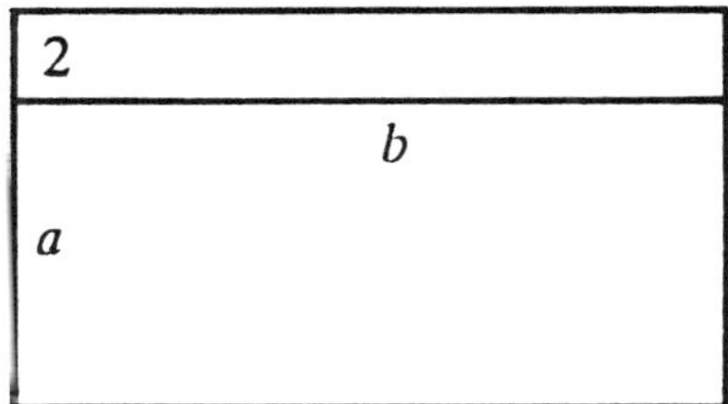

Figure 2.15

2.10. The answer is "yes". Indeed, if a is odd, then $a = 5 + 2k$ where $k \geq 0$ (we recall that $a \geq 5$). Consequently, the $a \times b$ rectangle can be decomposed into one $5 \times b$ block and several $2 \times b$ blocks (Figure 2.16). Thus the $a \times b$ rectangle can be tiled by L-trominoes (see Exercises 2.8 and 2.9).

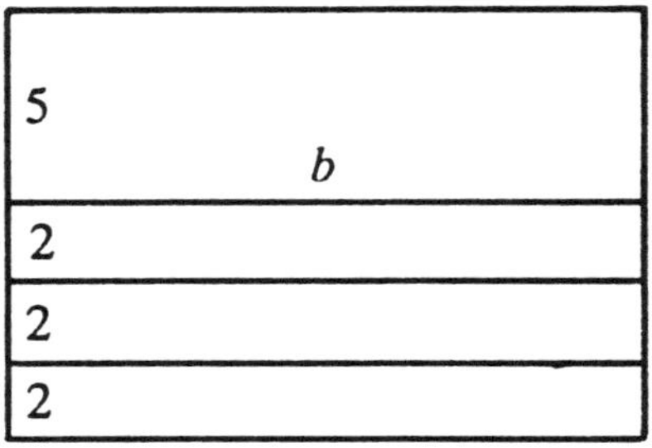

Figure 2.16

If a is even, then the $a \times b$ rectangle can be decomposed into several $2 \times b$ blocks (Figure 2.17) and, consequently, the $a \times b$ rectangle can be tiled by L-trominoes (see Exercise 2.3).

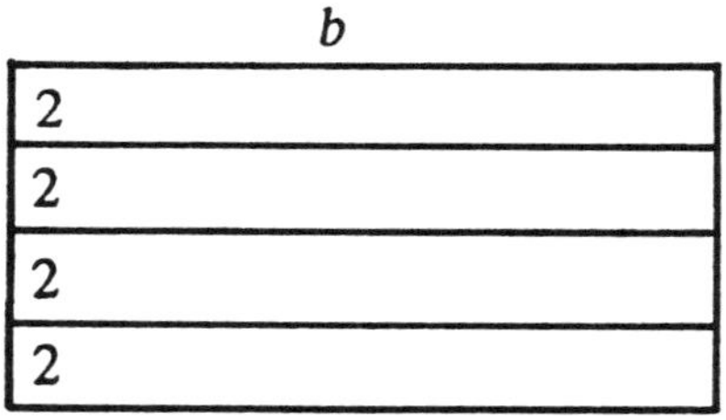

Figure 2.17

3. TETROMINOES AND "COLOR" REASONING

There exist five types of *tetrominoes* (Figure 3.1). We will call them O-, Z-, L-, T-, and I-tetrominoes. In this section we will consider the problem of tiling a rectangle by O-, Z-, L-, and T-tetrominoes. As to I-tetrominoes, the problem will be generalized and solved in the next section. In order to solve these problems we will often use the "color" reasoning which was mentioned in Section 1.

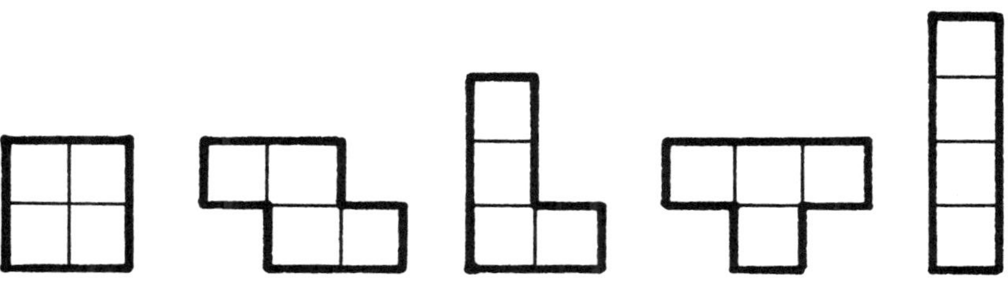

Figure 3.1

Exercise 3.1. Prove that an $m \times n$ rectangle can be tiled with O-tetrominoes if and only if both numbers m,n are even.

Exercise 3.2. Prove that there is no rectangle that can be tiled by Z-shaped tetrominoes.

The above exercises do not require the use of "color" reasoning. But for other types of tetrominoes such reasoning will be very useful.

Exercise 3.3. Prove that if all squares of an $a \times b$ chessboard are colored in two colors, black and white (Figure 3.2), then the difference between the number of black squares and the number of white squares is equal to either 0 or ± 1.

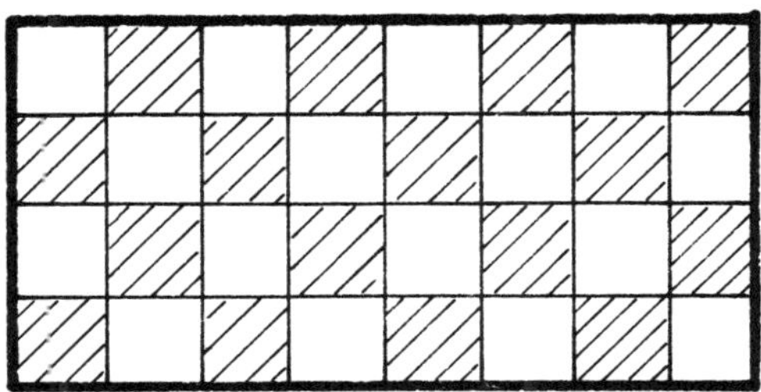

Figure 3.2

Exercise 3.4. Prove that if an $m \times n$ rectangle is colored in two colors, black and white by columns (Figure 3.3), then the difference between the number of black squares and the number of white squares is equal to either 0 or $\pm m$.

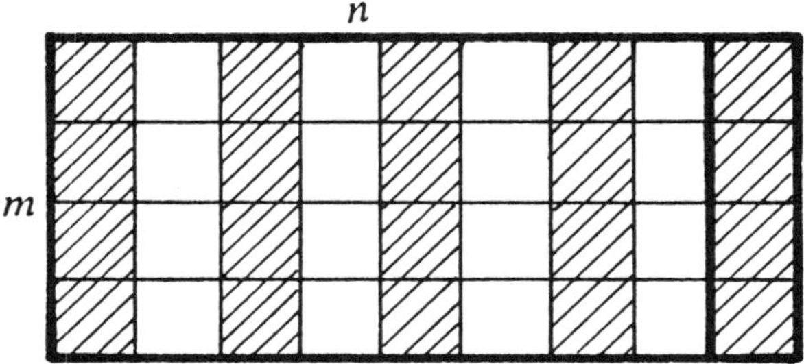

Figure 3.3

Exercise 3.5. Prove that if an $m \times n$ rectangle is tiled by L-tetrominoes, then the number of tiles is even.

This exercise shows that the area of a rectangle tiled by L-tetrominoes must be divisible by 8.

It is evident that the 2×4 rectangle can be tiled by two L-tetrominoes (Figure 3.4). The following problem arises naturally: *What other rectangles can be tiled by tetrominoes of this type?* Can you solve this problem yourself? If not, try to solve the following exercises that suggest a way of solving the indicated problem.

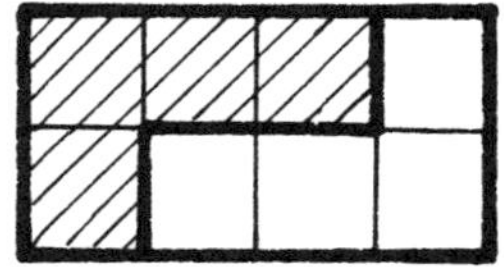

Figure 3.4

Exercise 3.6. Prove that each $m \times n$ rectangle, such that m is even and n is divisible by 4, can be tiled by L-tetrominoes.

Exercise 3.7. Prove that the 3×8 rectangle can be tiled by L-tetrominoes.

Exercise 3.8. ([GK] Prove that an $m \times n$ rectangle where $m \geq 2$ and $n \geq 2$ can be tiled by L-tetrominoes if and only if the product mn is divisible by 8.

The above exercise gives a complete solution of tiling a rectangle by L-tetrominoes. As to T-tetrominoes, solving a similar problem for them is much more difficult. First several steps we leave to the reader.

Exercise 3.9. Prove that if an $m \times n$ rectangle is tiled by T-tetrominoes, then the number of tiles is even.

Thus, the area of the rectangle that is tiled by T-tetrominoes is divisible by 8.

Exercise 3.10. Prove that the 4×4 square can be tiled by T-tetrominoes.

Exercise 3.11. Prove that the $n \times n$ square can be tiled by T-tetrominoes if and only if n is divisible by 4.

It is clear that an $m \times n$ rectangle where each of the numbers m, n is divisible by 4, can be tiled by T-tetrominoes (Figure 3.5; Exercise 3.10). This sufficient condition of tileability by T-shaped tetrominoes is also necessary. A clever proof of this fact was found by D.W. Walkup of Boeing Scientific Research Laboratories in 1965. Read it in the Monthly!

Theorem 3.1. (D.W. Walkup [Wa]]) *If an $m \times n$ rectangle can be tiled by T-tetrominoes, then m and n are integral multiples of 4.*

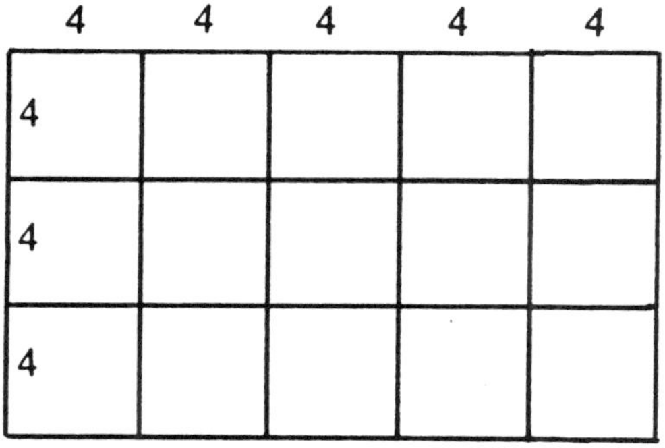

Figure 3.5

Solutions of Exercises

3.1. If both the numbers m,n are even, then the $m \times n$ rectangle can obviously be tiled (Figure 3.6).

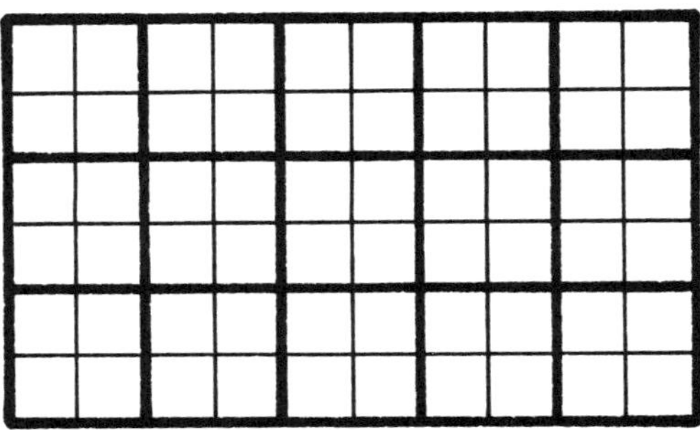

Figure 3.6

In order to prove the converse assertion, let us consider the coloring of the unit squares of rectangle shown in Figure 3.7. Each O-tetromino covers exactly *one* black square. Thus, if the rectangle is tiled by O-tetrominoes, then the number of tetrominoes in a row is equal to a where a is the number of the black squares in the upper row. Consequently, the number of the unit squares (black and white) in a row is equal to $2a$. Similarly, the number of the unit squares (black and white) in a column is equal to $2b$, where b is the number of black squares in the left column.

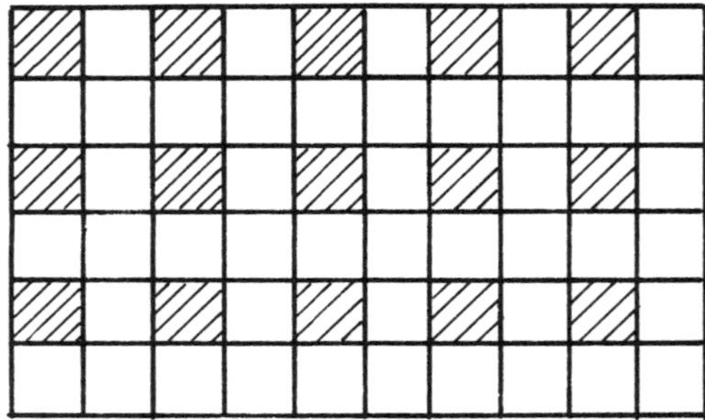

Figure 3.7

We can solve this exercise without coloring. Indeed, let a rectangle tiled by O-tetrominoes have the size $a \times b$ when the side of the tetromino is taken as the unit of measuring of lengths. If we return to the side of a unit square as the measuring unit, then the rectangle would have the size $2a \times 2b$.

3.2. There are two ways to cover the upper left square by a Z-tetromino (Figure 3.8). The two ways are symmetric, so it sufices to consider only the first one. Then, there is only one way to cover the third square in the upper row (Figure 3.9). Further, there is only one way to cover the fifth square in the upper row (Figure 3.10) and so on.

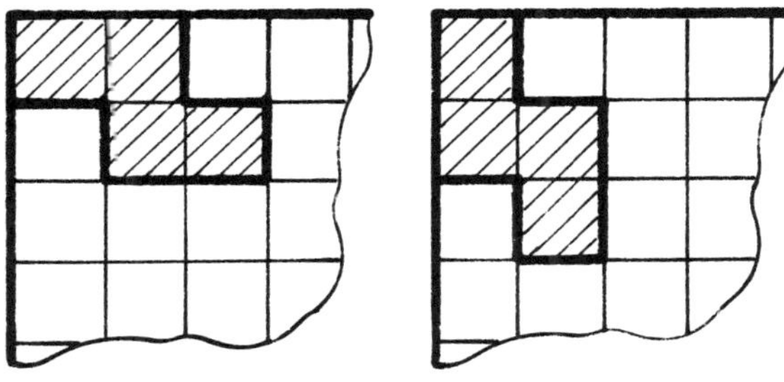

Figure 3.8

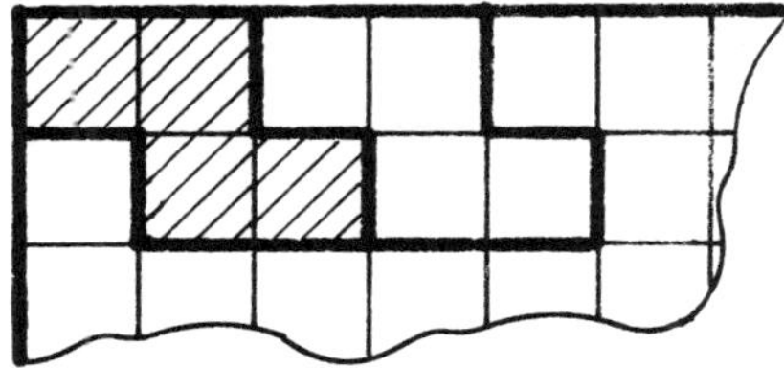

Figure 3.9

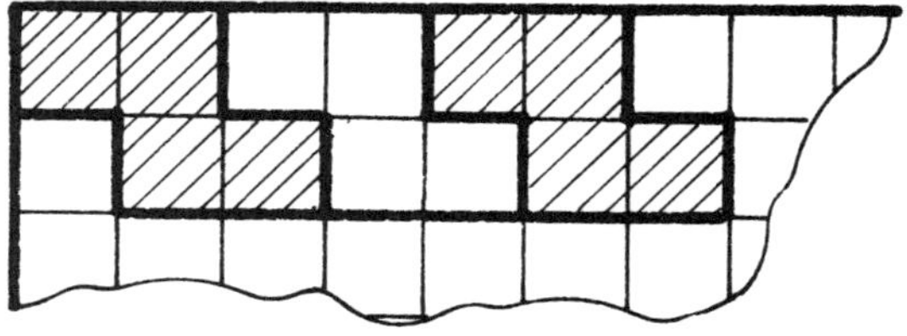

Figure 3.10

When we get to the right side of the rectangle (Figure 3.11), it would be impossible to cover the upper right corner (and stay inside the boundaries of the rectangle). Thus it is impossible to tile any rectangle by Z-tetrominoes.

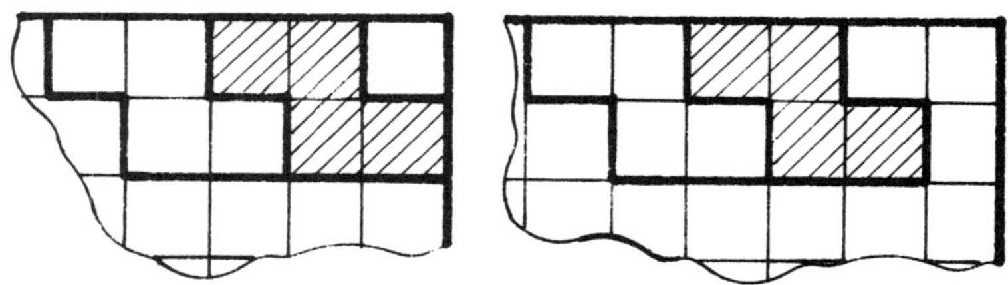

Figure 3.11

3.3. If the rectangle has an even number of columns, then the numbers of black and white squares are equal since the system of black squares is symmetric to the system of white squares with respect to the midline (Figure 3.12).

If the number of columns is odd, then the rectangle can be decomposed into a rectangle with even number of columns and a rectangle with only one column (Figure 3.13). But in a one-column rectangle the number of black squares is either equal to the number of white squares or differs from that number by one.

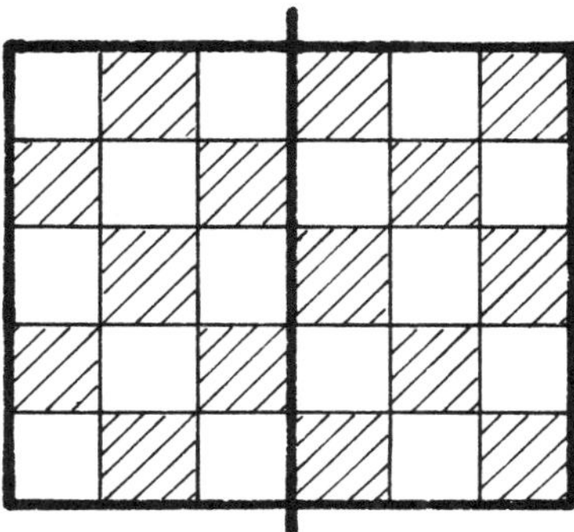

Figure 3.12

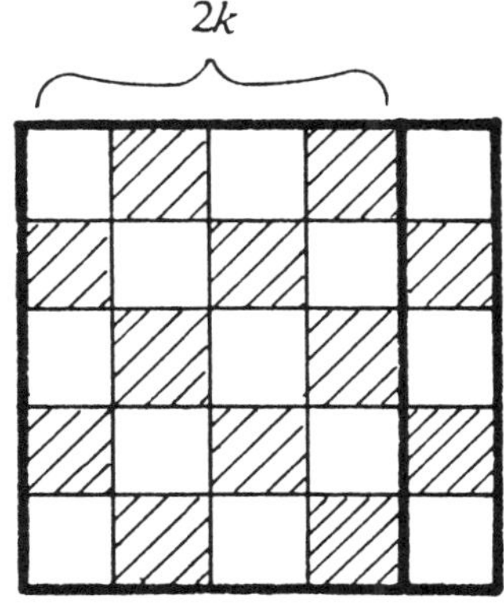

Figure 3.13

3.4. If the number of columns is even, then the numbers of black and white squares are equal. If the number of columns is odd, then there is one "extra" column. In this case the difference between the numbers of black and white squares is equal to $\pm m$ (Figure 3.3).

3.5. Assume that an $m \times n$ rectangle can be tiled by L-tetrominoes. Since the area of an L-tetromino is equal to 4, then $mn = 4k$ where k is the number of L-tetrominoes. Consequently, mn is divisible by 4, so at least one of the numbers m,n is even. Without loss of generality we can assume that n is even, that is the number of columns in the $m \times n$ rectangle is even.

22

Let us color the rectangle by alternating black and white columns. Each L-tetromino placed on the colored rectangle covers three unit squares of one color and one unit square of the other color (Figure 3.14). So we have "black" L-tetrominoes (three black and one white square) and "white" L-tetrominoes (three white and one black square). If the number p of "black" L-tetrominoes is not equal to the number q of "white" L-tetrominoes, then the numbers of black and white squares in the rectangle would not be equal. But they are equal since n is even; therefore, $p = q$, that is, the number $p + q$ of all L-tetrominoes is even. Therefore, the area mn of the rectangle is divisible by 8.

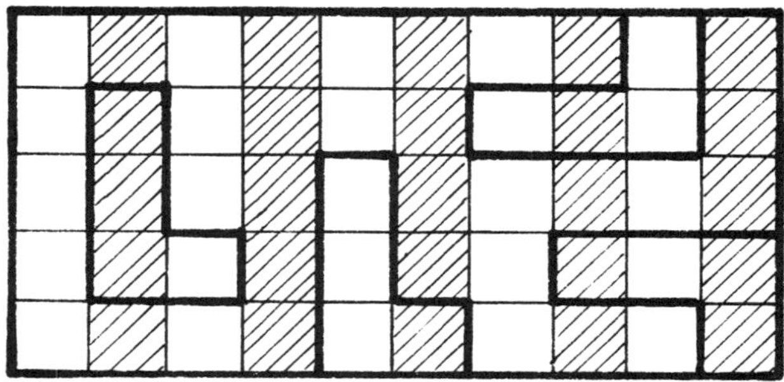

Figure 3.14

3.6. The $m \times n$ rectangle can be decomposed into 2×4 blocks, and consequently, it can be tiled by L-tetrominoes (see Figure 3.4).

3.7. A solution is shown in Figure 3.15. This can be said to be "essentially" the only solution. What does this mean?

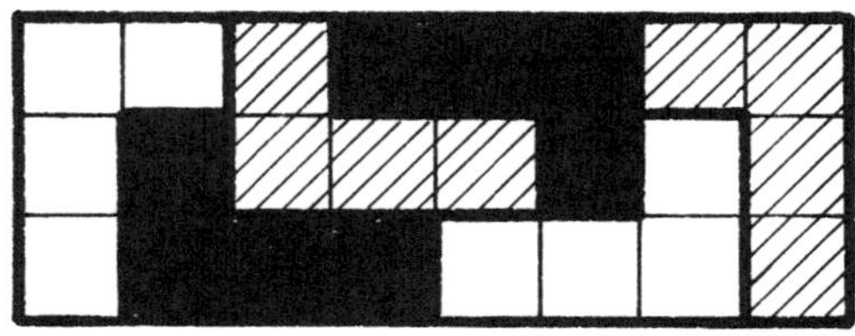

Figure 3.15

3.8. The indicated condition (divisibility by 8) is necessary. This follows from the solution of Exercise 3.5. Now we are going to prove that it is also sufficient.

Let M be an $m \times n$ rectangle, such that the product mn is divisible by 8. If both of the numbers m,n are even (and one of them is divisible by 4), then M can be tiled by L-tetrominoes (see Exercise 3.6). If one of the numbers a,b (say a) is odd, and consequently, the other is divisible by 8, then M can be decomposed into $a \times 8$ blocks (Figure 3.16). Further, since $a \geq 2$ is odd, then it is possible to decompose the $a \times 8$ block into the 3 $\times$ 8 block and an $(a - 3) \times 8$ block where $a - 3$ is even (Figure 3.17). Consequently (see Exercises 3.6 and 3.7) the $a \times 8$ block can be tiled by L-tetrominoes. Thus, the $a \times b$ rectangle can be tiled as well.

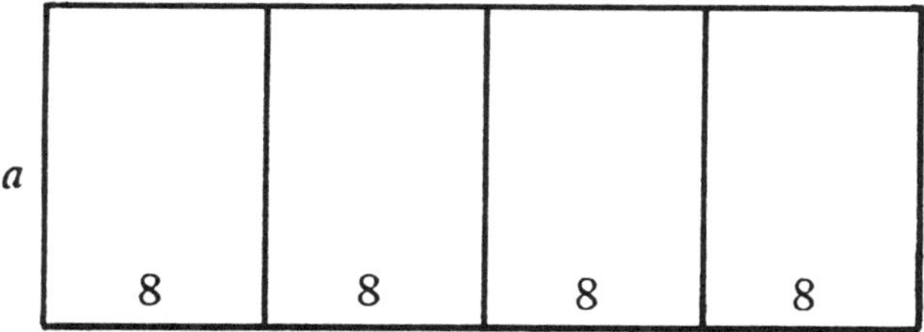

Figure 3.16

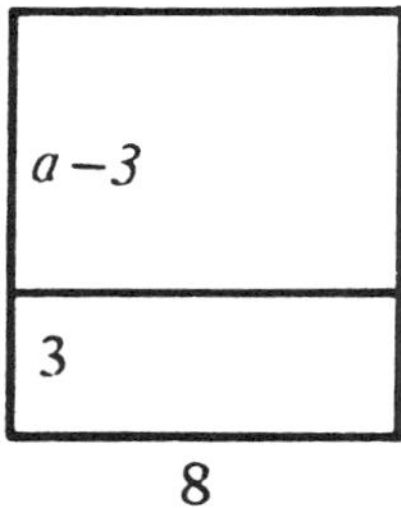

Figure 3.17

3.9. Consider the "chessboard" coloring of the rectangle. It can easily be shown that each T-tetromino placed on the rectangle covers three unit squares of one color and one unit square of the other color (Figure 3.18). So we have "black" T-tetrominoes (three black and one white square) and "white" T-tetrominoes (three white and one white square). If the number p of "black" T-tetrominoes were not equal to the number q of "white" T-tetrominoes, then the difference between the number of "black" squares and the number of "white" squares would be not less than 2. But this is impossible by virtue of the result of Exercise 3.3. So, $p = q$, that is, the number $p + q$ of all T-tetrominoes is even.

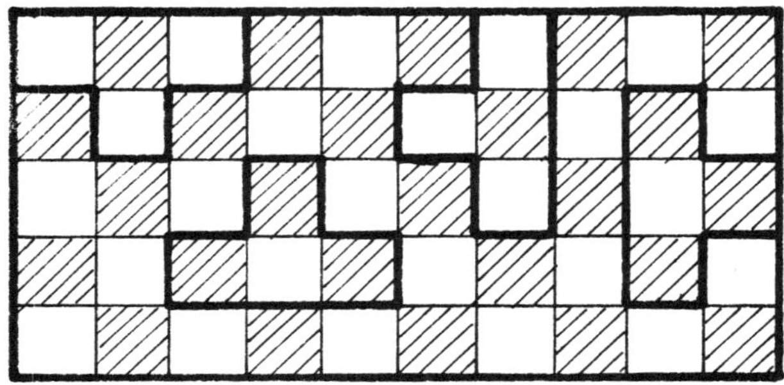

Figure 3.18

3.10. A solution is shown in Figure 3.19. Again, it is "essentially" the only solution.

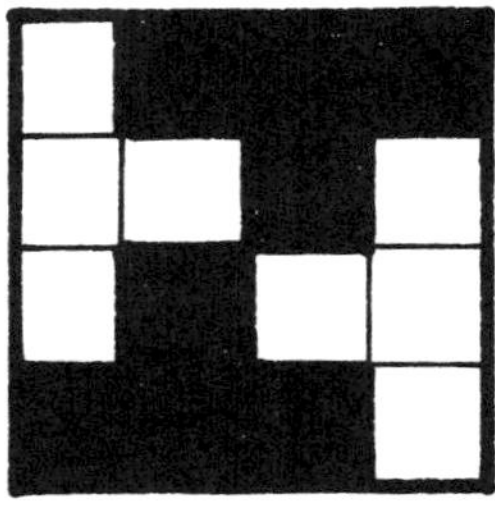

Figure 3.19

3.11. If the $n \times n$ square can be tiled by T-tetrominoes, then n^2 is divisible by 8 (see Exercise 3.8), and, consequently, n is divisible by 4. Conversely, if n is divisible by 4, then the $n \times n$ square can be decomposed into 4×4 blocks and hence the $n \times n$ square can be tiled by T-tetrominoes (see Exercise 3.10).

4. TILING BY LINEAR POLYOMINOES

In this section we are going to first solve the following problem.

Problem 4.1. Find all rectangles $m \times n$ that can be tiled by *linear k-ominoes* (i.e., $1 \times k$ rectangular tiles).

Exercise 4.1. If m or n is a multiple of k then the $m \times n$ board can be tiled by linear k-ominoes.

Theorem 4.1. (D.A. Klarner [Kl], A. Soifer [S3]) *If neither of m,n is a multiple of k then the m $\times$ n rectangle is not tileable by linear k-ominoes.*

First Proof. Assume that neither m nor n is a multiple of k, but that the board can be tiled by linear k-ominoes. Let us color the board diagonally in k colors with cyclic permutation of colors (see Figures 4.1 and 4.2).

This coloring (*diagonal cyclic k-coloring*) has a remarkable property: no matter how a linear k-omino is placed on the board, horizontally or vertically, it will cover exactly one square of each of the k colors.

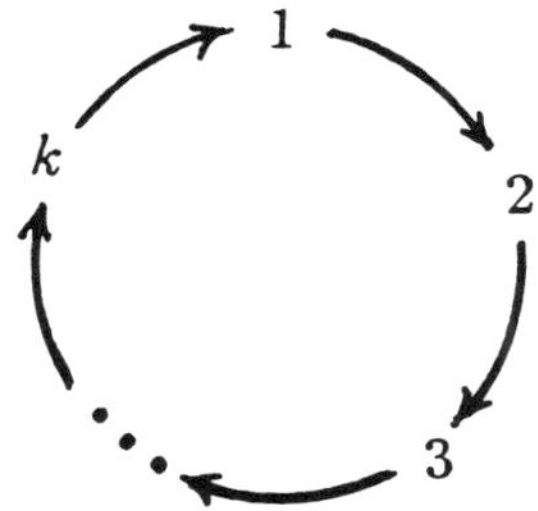

Figure 4.1

Diagonal cyclic k-coloring for $k = 3$

Figure 4.2

Since by assumption the board can be tiled by linear k-ominoes, and every k-omino covers an equal number of squares of every color, the board contains an equal number of squares of each of the k colors. It is not difficult to prove (do) that for any natural numbers m and k there are non-negative integers q and r_1, such that $0 \leq r_1 \leq \frac{k}{2}$ and

$$m = kq + r_1$$

or

$$m = kq - r_1$$

Accordingly, we will consider two cases:

Case 1. $m = kq + r_1$. We cut the given board into two rectangular boards: $kq \times n$ and $r_1 \times n$. Since the $kq \times n$ board can be tiled by linear k-ominoes, it contains an equal number of squares of each color. The given board contains an equal number of squares of every color as well, therefore the $r_1 \times n$ board has the same property.

Case 2. $m = kq - r_1$. In this case we attach the $r_1 \times n$ board to the given board to obtain the $kq \times n$ checker board and extend the coloring of the given board to a diagonal cyclic k-coloring of the $kq \times n$ board. The $kq \times n$ board can be tiled by linear k-ominoes, therefore it contains an equal number of squares of every color. The given $m \times n$ board contains an equal number of squares of every color, therefore the $r_1 \times n$ board contains an equal number of squares of every color.

Thus in both cases we got the $r_1 \times n$ board with diagonal cyclic k-coloring, which contains an equal number of squars of every color. Let us turn this board now by a 90° angle, i.e., consider the $n \times r_1$ board, and apply to it all of the above reasoning. As a result we will get an $r_2 \times r_1$ board, where $0 < r_2 \leq \frac{k}{2}$, that contains an equal number of squares of every color.

On the other hand, the number of one-color diagonals in the $r_2 \times r_1$ board is equal to $r_2 + r_1 - 1$, and

$$r_2 + r_1 - 1 \leq \frac{k}{2} + \frac{k}{2} - 1 < k$$

Therefore at least one of the k colors is not present at all in the $r_2 \times r_1$ board!

This contradiction proves that $m \times n$ board can be tiled by linear k-moinoes only if m or n is a multiple of k.

Second Proof. The following notation will be helpful for us: instead of writing "the remainders upon dividing numbers a_1, a_2, ... a_k by n are equal" we will simply write.

$$a_1 \equiv a_2 \equiv ... \equiv a_k (\text{mod } n)$$

*and say that a_1, a_2, ..., a_k are **congruent modulo n**.*

It is easy to prove (do) that $a_1 \equiv a_2 (\text{mod } n)$ if and only if $a_1 - a_2$ is a multiple of n.

Assume that neither of m, n is a multiple of k, i.e.,

$$m = kq_1 + r_1;\ 0 < r_1 < k$$

$$n = kq_2 + r_2;\ 0 < r_2 < k$$

but the board is tiled by linear k-ominoes.

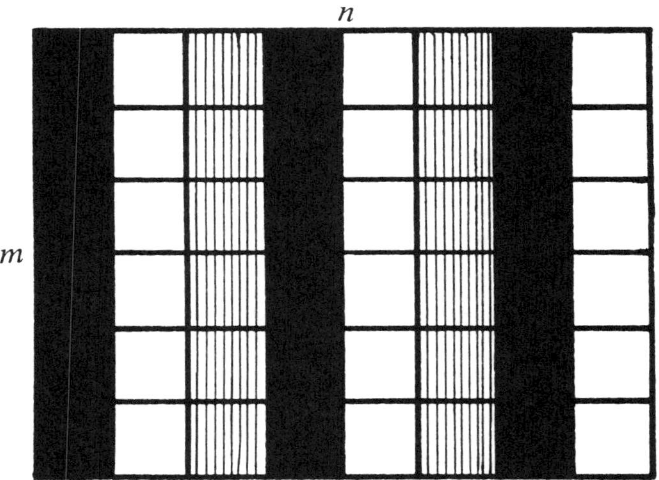

Column cyclic k-coloring for $k = 3$

Figure 4.3

Let us color the columns of the board each in one of the k colors with cyclic permutation of colors (see Figures 4.1 and 4.3) and denote by S_1, S_2, ..., S_n the numbers of squares of the board colored in the 1st, 2nd, ..., k-th colors respectively.

This coloring (***column cyclic k-coloring***) has an interesting property: if a linear k-omino is placed on the board vertically, it would cover k squares of the same color; if it is placed on the board horizontally, it would cover exactly one square of every color. By assumption the board is tiled by linear k-ominoes, therefore

$$S_1 \equiv S_2 \equiv \dots \equiv S_k \pmod{k}$$

On the other hand, you can notice that we have one column more of the r_2 - th color than of the $(r_2 + 1)$st color, i.e.,

$$S_{r_2} - S_{r_2 + 1} = m$$

Therefore

$$S_{r_2} \not\equiv S_{r_2 + 1} \pmod{k}$$

The contradiction we obtained proves that divisibility of m or n by k is a necessary condition for the $m \times n$ board to be tileable by linear k-ominoes. It is also sufficient (Exercise 4.1).

These solutions came from the book [S1].

Now to the unlimited collection of $1 \times k$ tiles let us add one, just one monomino.

Problem 4.2. (A. Soifer [S3]) What rectangles can be tiled by a combination of one monomino and linear k-ominoes? Where must we place the monomino?

Exercise 4.2. If an $m \times n$ rectangle is tiled by linear k-ominoes and one monomino, then $mn - 1$ is divisible by k.

We need a way to talk about a particular 1×1 square in an $m \times n$ rectangle without pointing a finger at the square. We can certainly number all unit squares of the rectangle: 1, 2, ..., mn. But it is a bulky way and not very convenient: it is tricky even to see whether two unit squares are neighbors or not. A much better way is to define on our $m \times n$ rectangle a cartesian coordinate system (Figure 4.4).

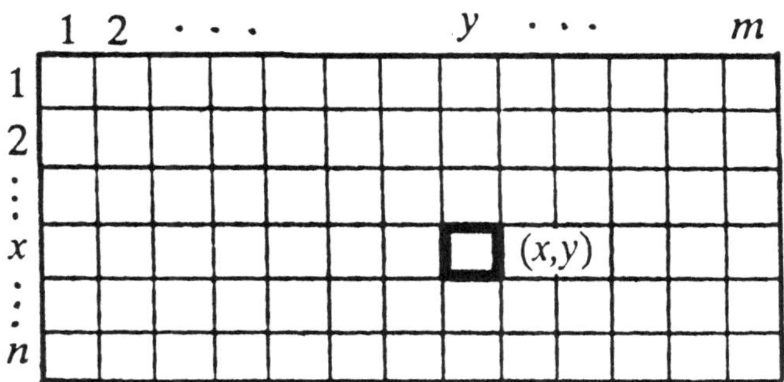

Figure 4.4

Each unit square is defined now by the ordered pair (x,y) of integers, where x and y are the number of the row and the number of the column in which the square is located.

Exercise 4.3. Solve Problem 4.2 for $k = 2$.

Now let us solve together the main part of Problem 4.2: $k > 2$.

Assume that an $m \times n$ rectangle is tiled by one monomino and linear k-ominoes. Let us color the rectangle in k colors by column cyclic coloring (please see Figures 4.1 and 4.3). As we've already noticed in the solution of Problem 4.1, this coloring has a nice property: if a linear k-omino is placed on the board vertically, it would cover k squares of the same color; if it is placed on the board horizontally, it would cover exactly one square of every color. By assumption the $m \times n$ board is tiled by linear k-ominoes and one unit square tile, therefore

$$S_1 \equiv S_2 \equiv \ldots \equiv S_k (\bmod k) \tag{1}$$

where S_i is the number of unit squares of color i in the given $m \times n$ rectangle without the square covered by the monomino ($i = 1, 2, \ldots, k$).

Let

$$m = kq_1 + r_1 \ (0 \le r_1 < k)$$
$$n = kq_2 + r_1 \ (0 \le r_2 < k)$$

Denote by P_i ($i = 1, 2, \ldots, k$) the number of unit squares of color i in the whole $m \times n$ rectangle.

Simple computations show that

$$P_1 = P_2 = \ldots = P_{r_1} = n \, (q_1 + 1) \tag{2a}$$
$$P_{r_1 + 1} = P_{r_1 + 2} = P_k = nq_1 \tag{2b}$$

Since the numbers $S_1, S_2, \ldots, S_k$ are all but one equal to the numbers $P_1, P_2, \ldots, P_k$ and one S_i is exactly one less than the corresponding P_i (we only covered one unit square by the monomino!), we have in view of (1) and (2) exactly two options:

Option 1. $r_1 = 1$ *and* the monomino is on color 1.

The monomino (let its coordinates be (x,y)) is on color 1, therefore

$$y \equiv 1(\text{mod } k)$$

Also,

$$m \equiv r_1 \equiv 1(\text{mod } k)$$

and

$$P_1 - 1 \equiv P_2 \equiv P_3 \equiv \dots \equiv P_k(\text{mod } k)$$

I.e.,

$$n(q_1 + 1) - 1 \equiv nq_1(\text{mod } k)$$

or

$$n \equiv 1(\text{mod } k)$$

Thus,

$$m \equiv n \equiv y \equiv 1(\text{mod } k)$$

Option 2. $r_1 = k - 1$ and the monomino is on color k.

The monomino (let its coordinates be (x,y)) is on color k, therefore

$$y \equiv 0(\text{mod } k)$$

In addition,

$$m \equiv r_1 \equiv -1(\text{mod } k)$$

and

$$P_1 \equiv P_2 \equiv \dots \equiv P_{k-1} \equiv P_{k-1}(\text{mod } k)$$

I.e.,

$$nq_1 - 1 \equiv n(q_1 + 1) \ (\text{mod } k)_1$$

therefore,

$$n \equiv -1(\text{mod } k)$$

Thus, in this case

$$m \equiv n \equiv -1(\text{mod } k)$$

and

$$y \equiv 0(\text{mod } k)$$

Now let us turn the $m \times n$ rectangle by 90°. I.e., consider the $n \times m$ rectangle and apply to it all the same reasoning. Our final conclusion: the following *must* take place

$$\begin{cases} m \equiv n \equiv 1 \pmod{k} \\ x \equiv y \equiv 1 \pmod{k} \end{cases} \quad \text{or} \quad \begin{cases} m \equiv n \equiv -1 \pmod{k} \\ x \equiv y \equiv 0 \pmod{k} \end{cases}$$

Problem 4.2 is half solved, however you still have to solve the following exercises.

Exercise 4.4. Prove that any $m \times n$ rectangle can be tiled by one monomino and linear k-ominoes if $m \equiv n \equiv 1 \pmod{k}$ and the coordinates (x,y) of the monomino satisfy the condition

$$x \equiv y \equiv 1 \pmod{k}$$

Exercise 4.5. Prove that any $m \times n$ rectangle can be tiled by one monomino and linear k-ominoes if $m \equiv n \equiv -1 \pmod{k}$ and the coordinates (x,y) of the monomino satisfy the condition

$$x \equiv y \equiv 0 \pmod{k}$$

Martin Gardner, who authored some two dozen magnificent books on popular mathematics, was one of the first readers of this manuspript. We are grateful to him for contributing the following new problem:

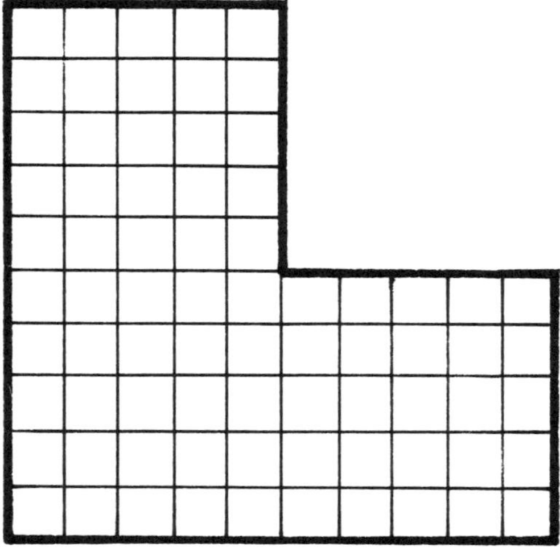

Figure 4.5

33

Exercise 4.6. (M. Gardner) Prove that the figure L above (a "giant L-tromino," see Figure 4.5) cannot be tiled with linear trominoes.

Solutions of Exercises

4.1. Behold:

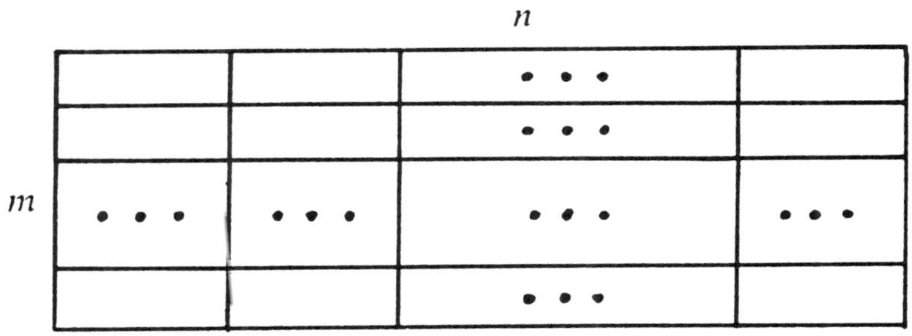

n

m

Figure 4.6

4.2. The area covered by linear k-ominoes is $mn-1$; it must be divisible by the area of one k-omino, i.e., by k.

4.3. Assume that the $m \times n$ board is tiled by one monomino located at (x,y) and dominoes. Then $mn - 1$ must be even; therefore, both m and n must be odd. Color the board in a chessboard fashion in black and white colors (black in the corner $(1,1)$), and observe that every domino covers exactly one square of each of the colors. Since the board has one more black square than whites the monomino must be on a black square, i.e.,

$$m \equiv n \equiv 1(\mathrm{mod}\ 2) \tag{3a}$$

and

$$x \equiv y(\mathrm{mod}\ 2) \tag{3b}$$

All there is left to show is that conditions (3) are also sufficient. Figures 4.7 and 4.8 do just that. Behold!

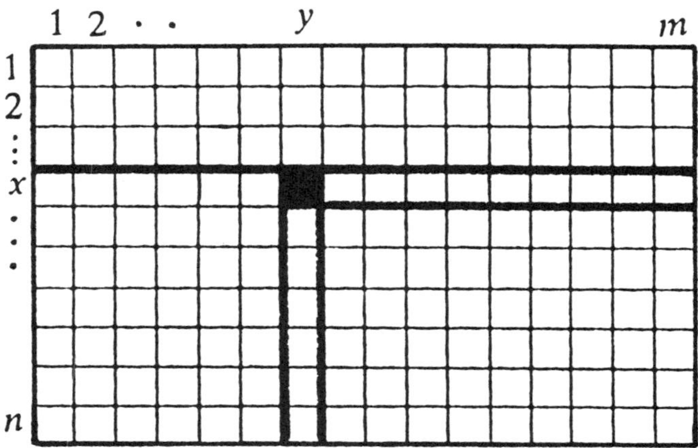

$$m \equiv n \equiv 1(\mathrm{mod}\ 2)$$
$$x \equiv y \equiv 1(\mathrm{mod}\ 2)$$

Figure 4.7

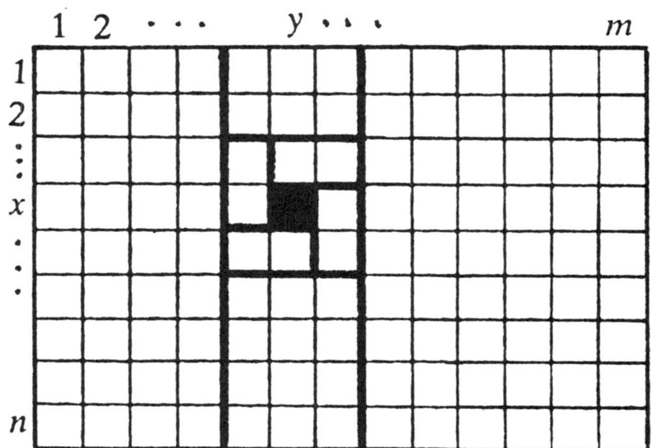

$$m \equiv n \equiv 1(\mathrm{mod}\ 2)$$
$$x \equiv y \equiv 0(\mathrm{mod}\ 2)$$

Figure 4.8

4.4. See:

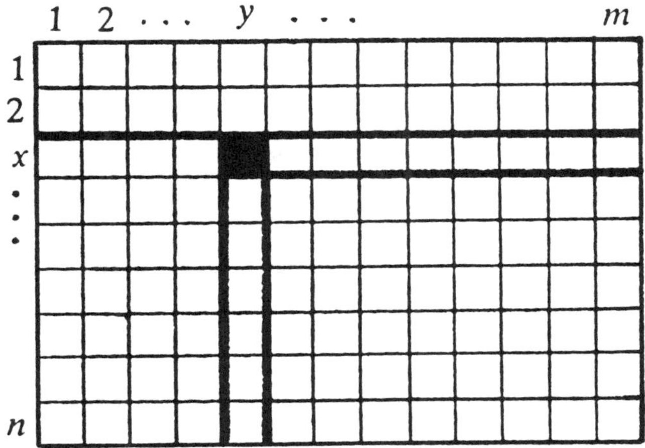

$$m \equiv n \equiv 1 \pmod{k}$$
$$x \equiv y \equiv 1 \pmod{k}$$

Figure 4.9

4.5. See:

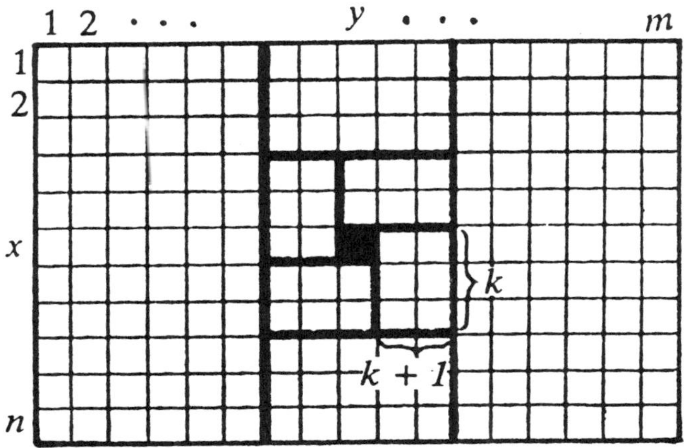

$$m \equiv n \equiv 1 \pmod{k}$$
$$x \equiv y \equiv 0 \pmod{k}$$

Figure 4.10

4.6. Let us color the figure L in three colors by column cyclic coloring (Figure 4.11). If a linear tromino is placed on L horizontally, it would cover exactly one square of every color; if a linear tromino is placed on L vertically, it would cover three squares of the same color.

Assume that the figure L is tiled by linear trominoes, then

$$S_1 \equiv S_2 \equiv S_3 \ (\text{mod } 3) \tag{4}$$

where S_1, S_2, S_3 are the numbers of black, grey, and white unit squares respectively.

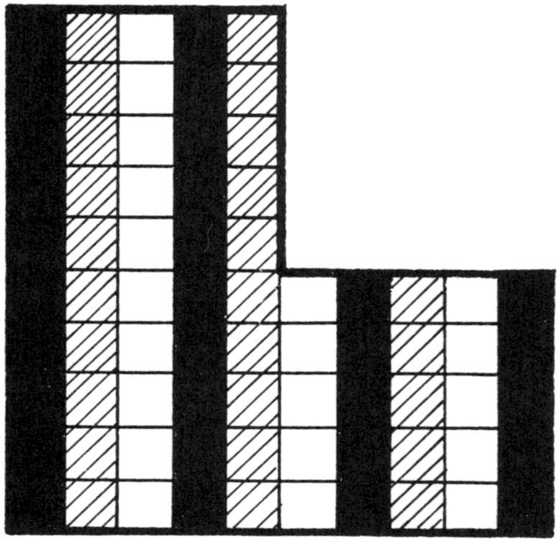

Figure 4.11

On the other hand, $S_1 = 30$; $S_2 = 25$; $S_3 = 20$ which contradicts to (4) above. We are done.

5. POLYOMINOES AND ROTATIONAL SYMMETRIES

A figure M is said to possess **n-fold rotational symmetry** if there exists a point c (the *center* of the symmetry or, more precisely, the n-fold rotational center), such that M is mapped onto itself under the rotation by $\frac{360°}{n}$ about c. For example, an arbitrary parallelogram possesses 2-fold rotational symmetry i.e., it is *centrally symmetric* (Figure 5.1). Each square is a figure with 4-fold rotational symmetry (Figure 5.2). The polygons in Figure 5.3 possess 3-fold rotational symmetry. The regular n-gon is a figure with n-fold rotational symmetry.

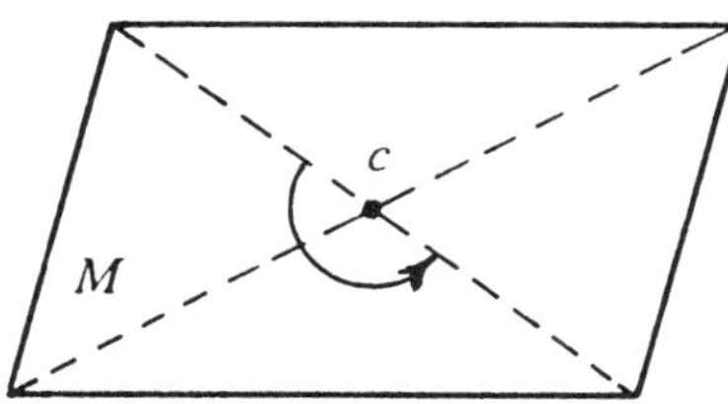

Figure 5.1

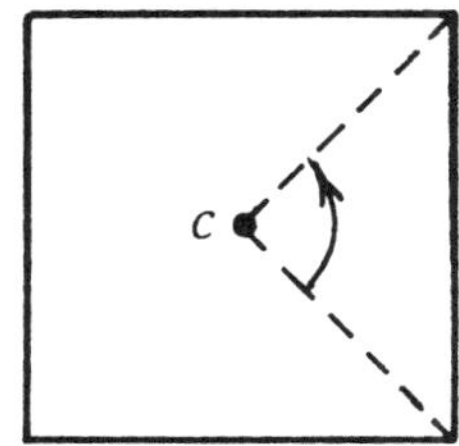

Figure 5.2

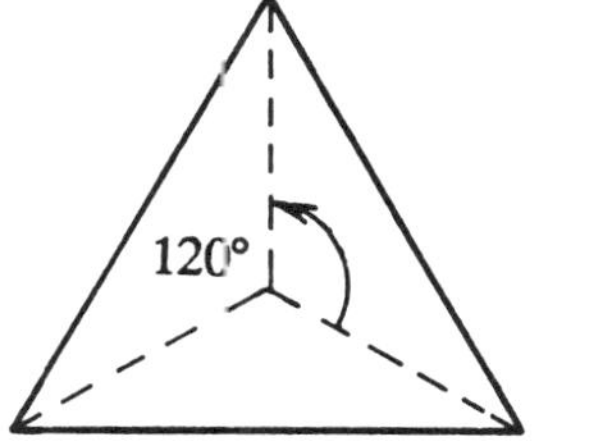

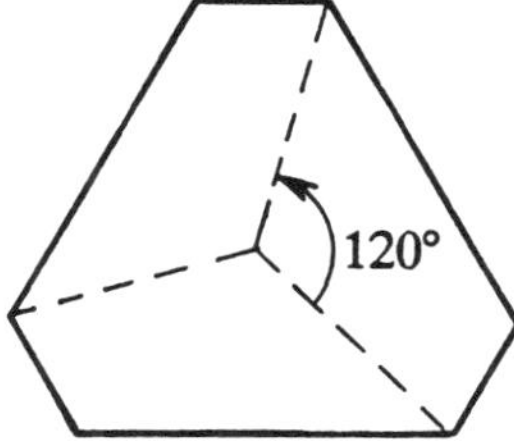

Figure 5.3

Now let us take a look at a rectangle tiled by polyominoes. We would say that this *tiling T has **n-fold rotational symmetry** if there is a point c such that T is mapped onto itself under the rotation by $\frac{360°}{n}$ about c.*

In Figures 2.2 and 3.4 the tilings are centrally symmetric, i.e., they possess 2-fold rotational symmetry. In Figure 3.19 the tiling of the 4×4 square by T-tetrominoes has 4-fold rotational symmetry.

Certainly no rectangle M possesses n-fold rotational symmetry except in the case $n = 2$ (and $n = 4$ if M is a square). So, the problem arises: *in what cases does there exist a tiling of a rectangle (or a square) by polyominoes with 2-fold (or 4-fold) rotational symmetry.*

Exercise 5.1. Prove that if both the numbers m,n are odd then there exists no centrally symmetric tiling of the $m \times n$ rectangle by L-trominoes.

Exercise 5.2. Prove that if the number m is even and the number n is divisible by 3, then the $m \times n$ rectangle possesses a centrally symmetric tiling by L-trominoes.

Exercise 5.3. Prove that if n is divisible by 6 then the $5 \times n$ rectangle has a centrally symmetric tiling by L-trominoes.

Exercise 5.4. Prove that if the number n is divisible by 6, then (for arbitrary $m \geq 2$) the $m \times n$ rectangle possesses a centrally symmetric tiling by L-trominoes.

Combining the results of Exercises 5.1-5.4 we obtain the following assertion:

Theorem 5.1. *The $m \times n$ rectangle possesses a centrally symmetric tiling by L-trominoes, if and only if the product mn is divisible by 6.*

In the next exercise we will consider tilings by L-trominoes with 4-fold rotational symmetry. Of course, such tilings exist only for squares.

Exercise 5.5. Prove that there exists a centrally symmetric tiling of the 6×6 square by L-trominoes.

Exercise 5.6. Prove that the $n \times n$ square possesses a tiling by L-trominoes with 4-fold rotational symmetry, if and only if n is divisible by 6.

The above propositions give a complete solution of the problem on tiling of rectangles by L-trominoes with rotational symmetries. Now we are going to look into tilings by L- and T-tetrominoes.

Exercise 5.7. Prove that if the number m is even and the number n is divisible by 4, then there exists a centrally symmetric tiling of the $m \times n$ rectangle by L-tetrominoes.

Exercise 5.8. Prove that there is no centrally symmetric tiling of the 3×8 rectangle by L-tetrominoes, but there exists such a tiling for the 3×16 rectangle.

Exercise 5.9. Prove that the $n \times n$ square possesses a tiling by L-tetrominoes with 4-fold rotational symmetry if and only if n is divisible by 4.

Exercise 5.10. Prove that the $n \times n$ square possesses a tiling by T-tetrominoes with 4-fold rotational symmetry if and only if n is divisible by 4.

There is probably no centrally symmetric tiling of the $m \times 8$ rectangle by the L-tetrominoes for any odd m, but the authors don't have a proof of this conjecture.

Finally, we would like to pose some interesting open problems.

Problem 5.1. Let M be a polyomino and P be a rectangle tiled by copies of M. We say that P is *minimal*, if P cannot be decomposed into two or more rectangles, each of which can be tiled by M. For example, the 2×3 and 5×9 rectangles are minimal for the L-tromino (and there are no other minimal rectangles). The question is: *is there for each given number p a polyomino M, such that there exists more than p minimal rectangles for this polyomino?*

Problem 5.2. A polyomino M consisting of k^2 unit squares has the $2k \times 2k$ square as its minimal rectangle. Is it true then that any tiling of the $2k \times 2k$ square by M possesses 4-fold rotational symmetry?

Two L-tetrominoes are said to have the same **orientation** *if there is a rotation that maps one of the tetrominoes onto the other.*

The existence of two orientations of L-tetrominoes can be illustrated by the following exercise.

Exercise 5.11. One side of the 4×6 rectangular piece R of paper is painted red, and the other is left white. Show that R can be cut into L-tetrominoes in such a way that one can not put them together to form a 3×8 rectangle with one side completely red.

And now an open problem:

Problem 5.3. Is it true that an $m \times n$ rectangle is tileable by L-tetrominoes of the same orientation if and only if one of the numbers m,n is divisible by 4 and the other is even?

Solutions of Exercises

5.1. If both numbers m,n are odd then there is a *central unit square* in the $m \times n$ rectangle. The center of this square coincides with the center of the whole rectangle (Figure 5.4). If we

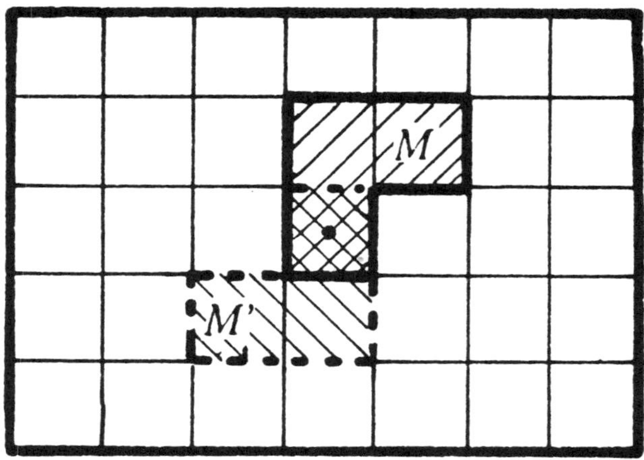

Figure 5.4

consider an *L*-tromino *M* that contains the central square, then we find that the symmetric *L*-tromino *M' overlaps M*. This shows that there exists no centrally symmetric tiling by *L*-trominoes.

5.2. Let us decompose the $m \times n$ rectangle into 2×3 blocks, as shown in Figures 5.5, 5.6, and 5.7.

If $m = 2k$ where k is even (as in Figure 5.5), then we can divide the rectangle into two $k \times n$ rectangles, each of which is decomposed into 2×3 blocks. So, constructing a tiling of the upper rectangle by *L*-trominoes (see Figure 2.2) and taking the symmetric tiling (with respect to the center of the whole rectangle) of the lower rectangle, we obtain a centrally symmetric tiling of the whole $m \times n$ rectangle.

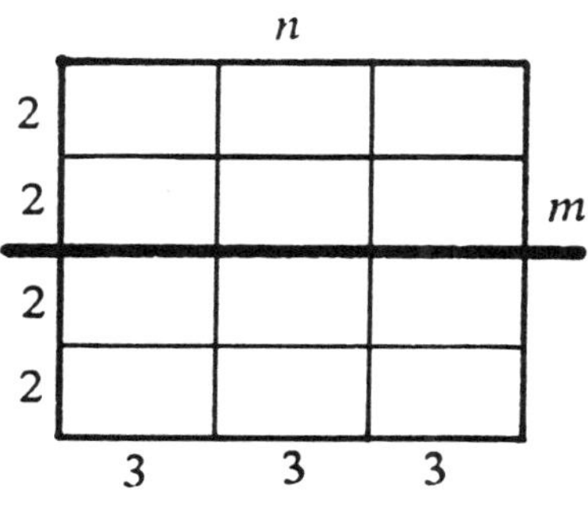

Figure 5.5

If $n = 3e$, where the number e is even (Figure 5.6), then the construction is similar.

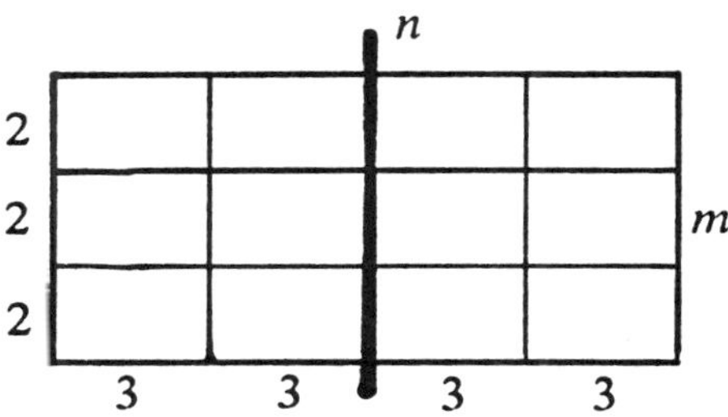

Figure 5.6

Finally let us consider the case when $m = 2k$, $n = 3e$ where both the numbers k, e are odd. Then there exists the *central* 2×3 block M in the $m \times n$ rectangle (Figure 5.7), and for any other 2×3 block P there exists the symmetric block P'. So, choosing a tiling of the central 2×3 block M by L-trominoes (it is centrally symmetric, see Figure 2.2) and choosing for each two symmetric 2×3 blocks P, P' respectively symmetric tilings, we obtain a centrally symmetric tiling of the whole rectangle.

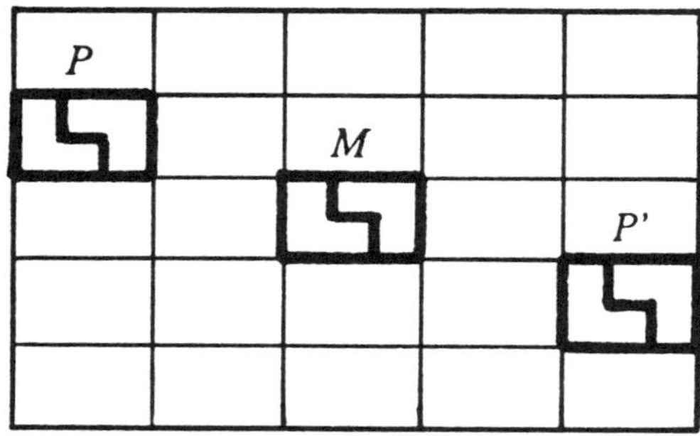

Figure 5.7

5.3. A tiling for the 5×6 rectangle is shown in Figure 5.8.

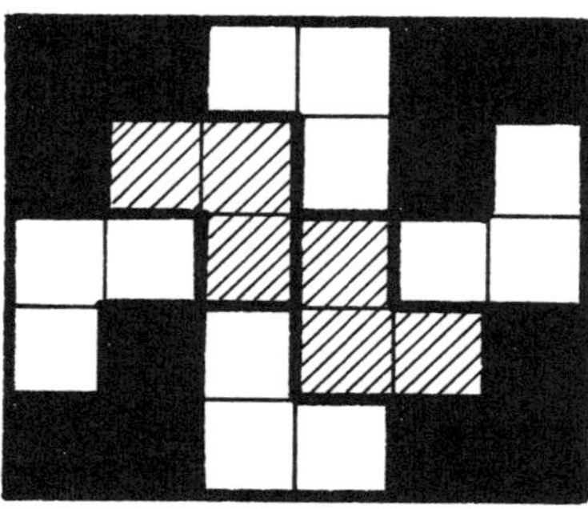

Figure 5.8

Further, for the $5 \times 6k$ rectangle we must consider separately the cases when k is even or odd. They are trivial, however (Figure 5.9).

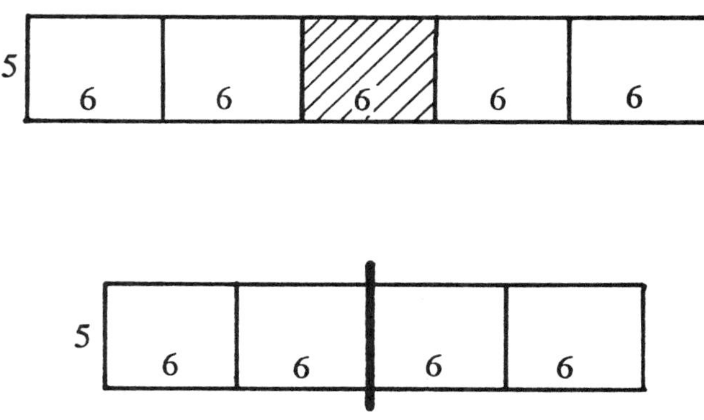

Figure 5.9

5.4. If m is even, then the required assertion follows directly from the result of Exercise 5.2. Let m now be odd. Then either $m = 4k + 3$, or $m = 4k + 5$, where $k \geq 0$. Therefore, we can decompose the $m \times n$ rectangle either into one central $3 \times n$ block and two $2k \times n$ blocks (Figure 5.10) or into one central $5 \times n$ block and two $2k \times n$ blocks (Figure 5.11). This gives a solution (see Exercises 5.2, 5.3, and 5.4), since n is divisible by 6.

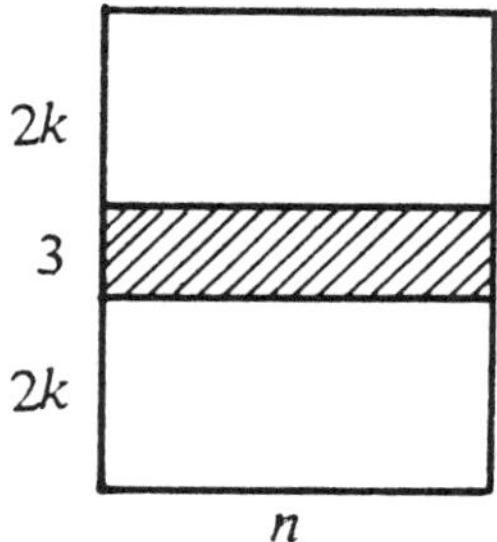

Figure 5.10

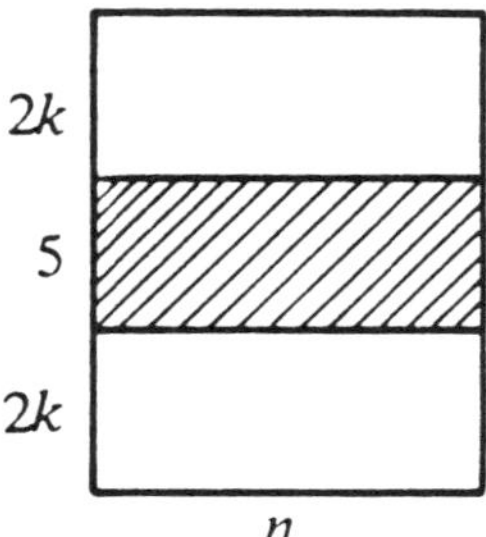

Figure 5.11

5.5. A solution is shown in Figure 5.12. It can be obtained in the following way. We take four L-trominoes at the corners of the 6 × 6 square and the tiling of the central 4 × 4 square by T-tetrominoes (Figure 5.13). This tiling (by four dominoes, four L-trominoes, and four T-tetrominoes) possesses 4-fold rotational symmetry. In order to receive the tiling of Figure 5.12, it remains now to "correct" the tiling in Figure 5.13 by replacing the union of one domino and one tetromino (shaded in Figure 5.13) by two L-trominoes.

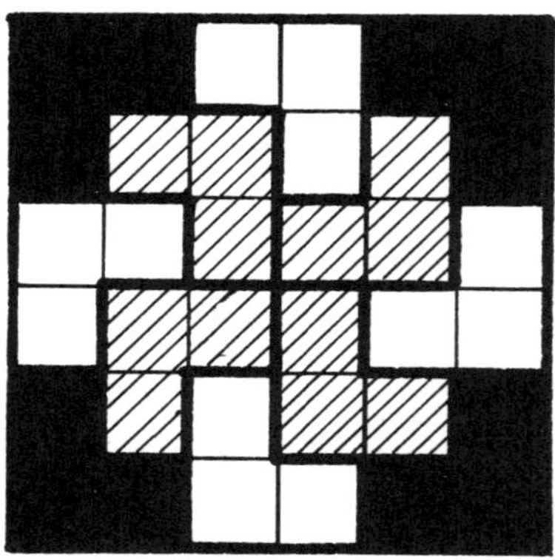

Figure 5.12

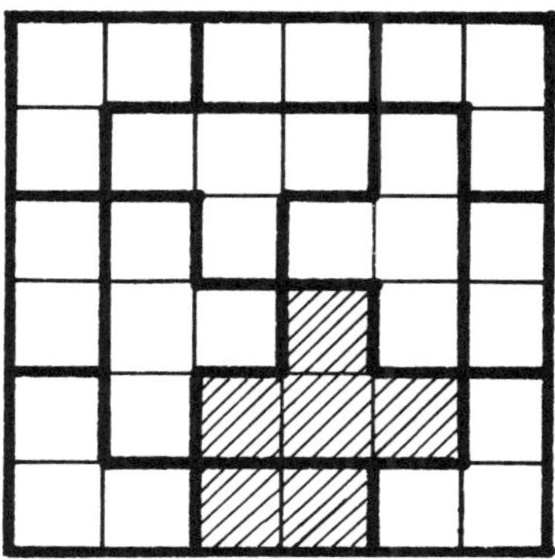

Figure 5.13

5.6. By virtue of the result of Theorem 5.1, if the $n \times n$ square possesses a tiling by L-trominoes with 4-fold rotational symmetry (and consequently, with 2-fold rotational symmetry), then the number n^2 is divisible by 6. Therefore, n is divisible by 6.

Conversely, if n is divisible by 6, that is, $n = 6k$, then the $n \times n$ square can be decomposed into 6×6 blocks. Now (separately for even and odd k, see Figures 5.14 and 5.15), we can construct a tiling of the $n \times n$ square by L-trominoes with 4-fold rotational symmetry (see Figure 5.12).

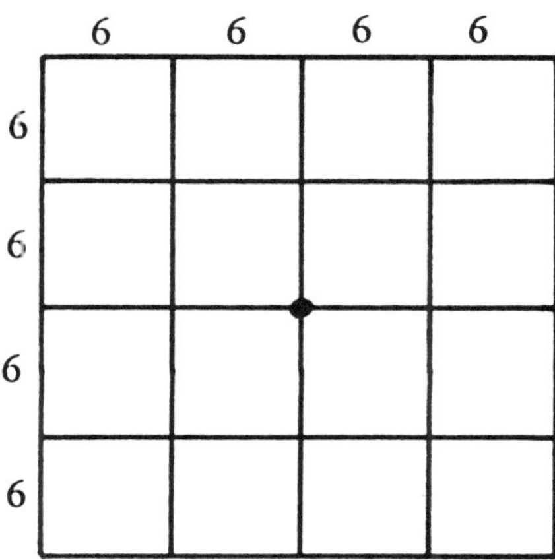

Figure 5.14

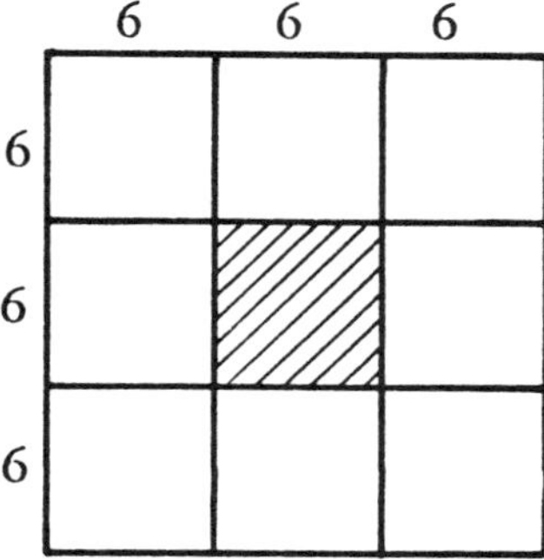

Figure 5.15

5.7. Let $m = 2k$ and $n = 4e$. If k is even, then we can decompose the $m \times n$ rectangle into 2×4 blocks in a way that is centrally symmetric (Figure 5.16). Figure 3.4 shows how to complete the required tiling. Similarly, if e is even, we can construct a tiling required (Figure 5.17).

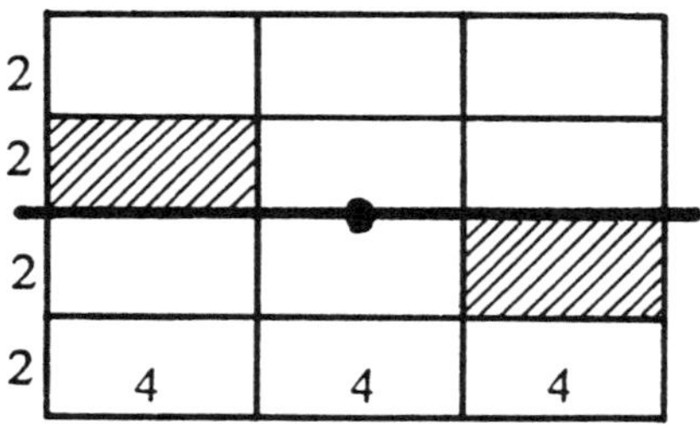

Figure 5.16

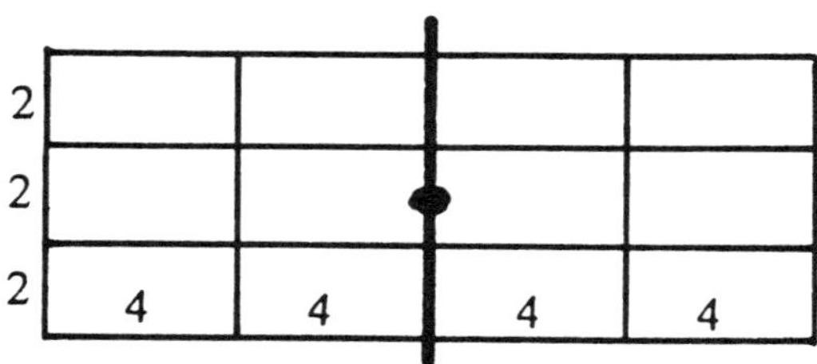

Figure 5.17

Finally, if both the numbers k, e are odd, then in the decomposition into 2×4 blocks there is the *central* block (Figure 5.18), and for each non-central 2×4 block P there is the symmetric block P'. This allows us to construct a centrally symmetric tiling of the rectangle by the L-tetrominoes (see the solution of Exercise 5.2).

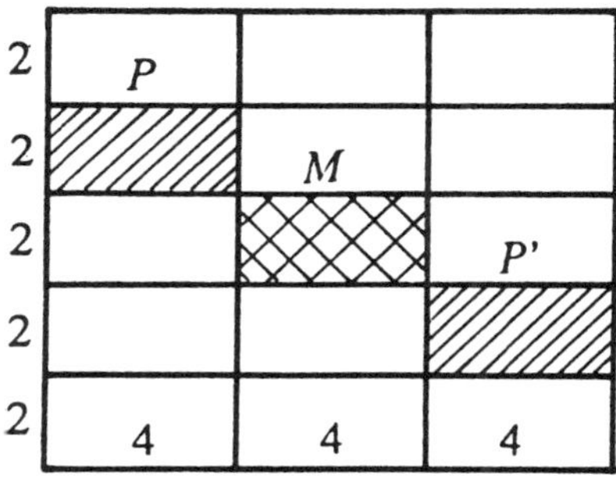

Figure 5.18

5.8. There are three possibilities to cover the left upper corner of the 3×8 rectangle (in Figure 5.19 each time with the T-tetromino we drew its symmetric one).

In Figure 5.19a we have two possible continuations (Figures 5.20a and 5.20b), but none of them allows the completion of tiling.

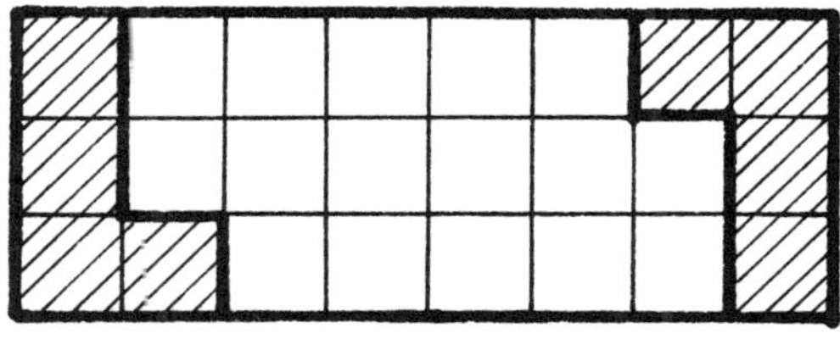

Figure 5.19a

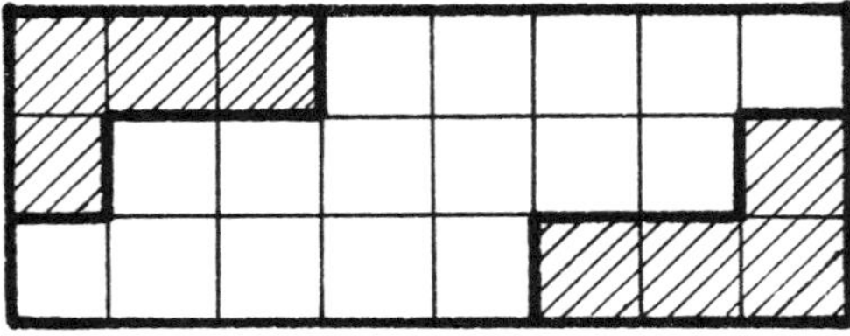

Figure 5.19b

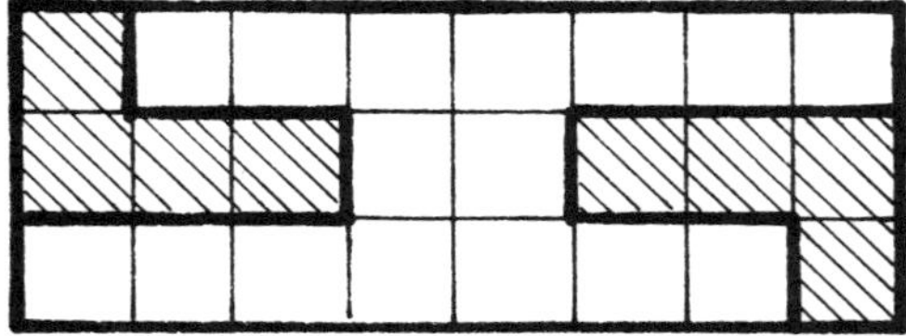

Figure 5.19c

In Figure 5.19b there is only one way to cover the left bottom corner (Figure 5.21) and it does not allow the completion of tiling either.

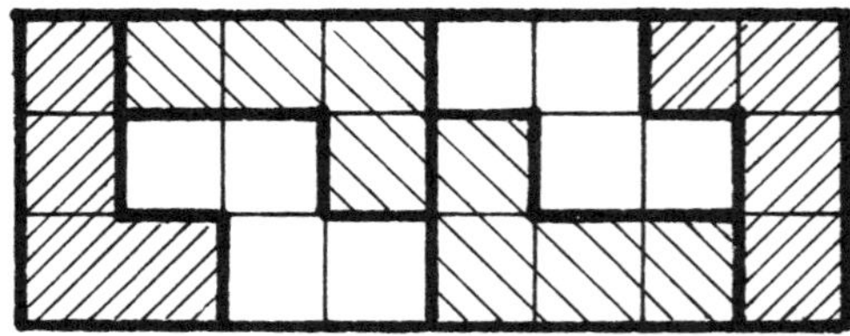

Figure 5.20a

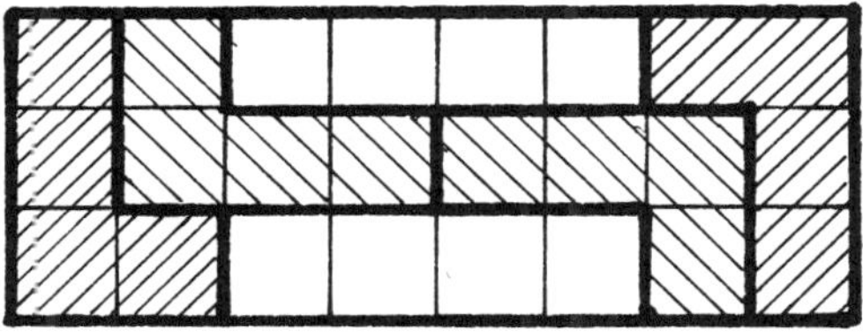

Figure 5.20b

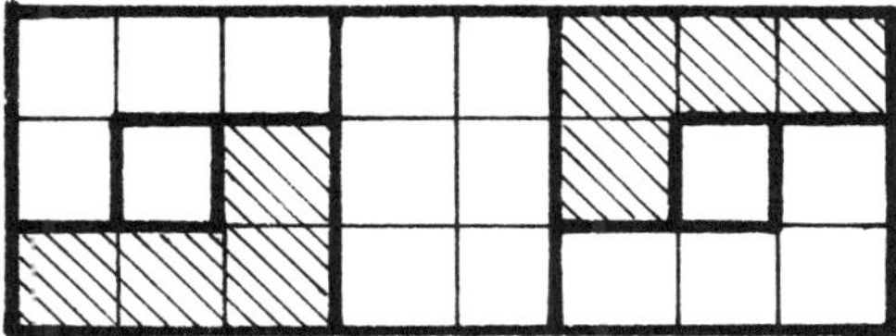

Figure 5.21

Finally in Figure 5.19c we have no way to cover the lower left corner. So, there exists no centrally symmetric tiling of the 3 × 8 rectangle by L-tetrominoes.

As to a 3 × 16 rectangle, we can take a tiling of the 3 × 8 rectangle in the left half of Figure 5.22, and then construct the tiling, symmetric to the former one with respect to point c in the right part of Figure 5.22. Thus we obtain a centrally symmetric tiling of the 3 × 16 rectangle by L-tetrominoes.

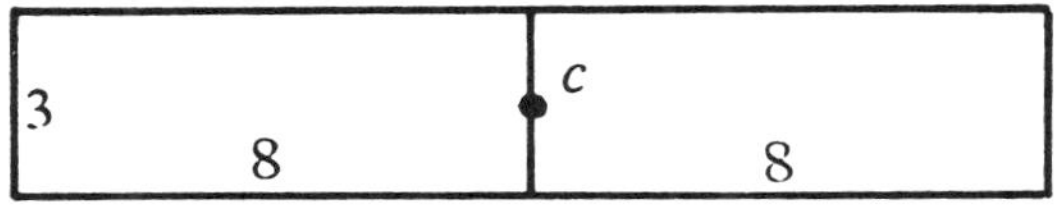

Figure 5.22

5.9. It follows from Theorem 3.2 of section 3 that if the $n \times n$ square possesses a required tiling, then the number n^2 is divisible by 8. Consequently, n is divisible by 4.

Conversely, if the number n is divisible by 4, then the $n \times n$ square can be decomposed into 4×4 blocks, and consequently, this square possesses a centrally symmetric tiling by L-tetrominoes.

5.10. The solution is the same as for Exercise 5.9.

5.11. Cut R ito L-tetrominoes of the same orientation. Then show that the 3×8 rectangle cannot be tiled by L-tetrominoes of the same orientation.

6. TILING ON OTHER SURFACES

In the preceding sections we were tiling an $m \times n$ rectangle R. We can glue together two opposite sides of R and thus form a *cylinder* $C(m \times n)$. In fact, we can go further and glue together the two bases of the cylinder and thus form a donut (or bagel, depending upon your culinary taste). Mathematicians call it *torus* (Figure 6.1), so we will denote it by $T(m \times n)$.

Of course, if the $m \times n$ rectangle is tileable by, say, tiles of shape T, so will be the cylinder $C(m \times n)$. And if the cylinder $C(m \times n)$ is tileable, so will be the torus $T(m \times n)$. But the converse is not always true. Sometimes cylinders are "more tileable" than the original rectangle, and in turn, the torus can be "more tileable" than the cylinder. Please also note that the $m \times n$ rectangle gives birth to two cylinders, $C(m \times n)$ and $C(n \times m)$ depending upon which pair of opposite sides is glued together. We, however, get only one torus $T(m \times n)$. Indeed, in order to obtain the torus $T(m,n)$ from the original $m \times n$ rectangle directly, we can glue together both pairs of the opposite sides of the rectangle simultaneoulsy.

Now your are ready for explorations on your own!

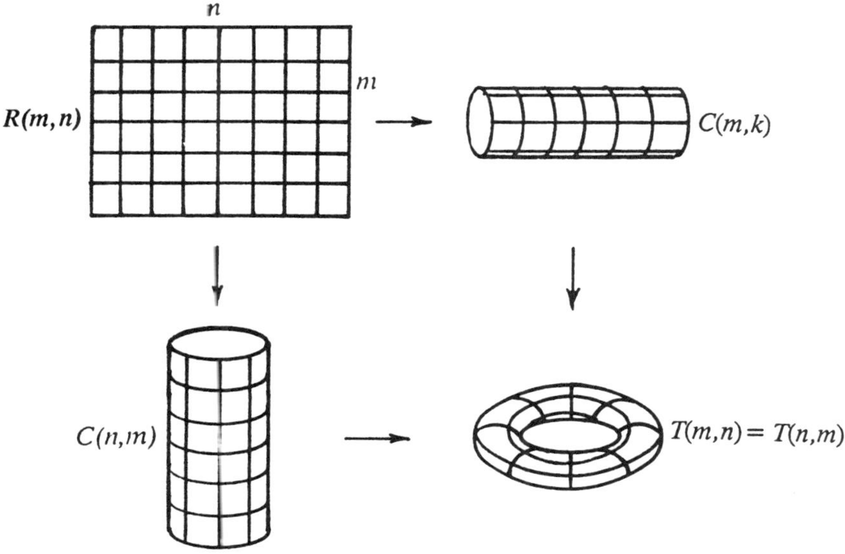

Figure 6.1

Exercise 6.1. Find all n such that the torus $T(n \times n)$ can be tiled by L-trominoes.

Exercise 6.2. Can it so happen that the cylinder $C(m \times n)$ is tileable by T-tetrominoes, where as the cylinder $C(n \times m)$ is not?

Exercise 6.3. Find all m and n, such that the cylinder $C(m \times n)$ can be tiled by linear k-ominoes. Can it so happen that exactly one of the cylinders $C(m \times n)$, $C(n \times m)$ can be tiled by linear k-ominoes?

Let us go back to our original $m \times n$ rectangle. If we first twist the rectangle and then glue its opposite sides, we get what is known as the **Möbius band** $M(m \times n)$ (Figure 6.2). We imagine this surface, as a mathematical abstraction, without thickness. It is a very unusual surface. If you travel from any point along the

52

Möbius band you will get back to P but you will be situated on the other side of the Möbius band (Figure 6.3).

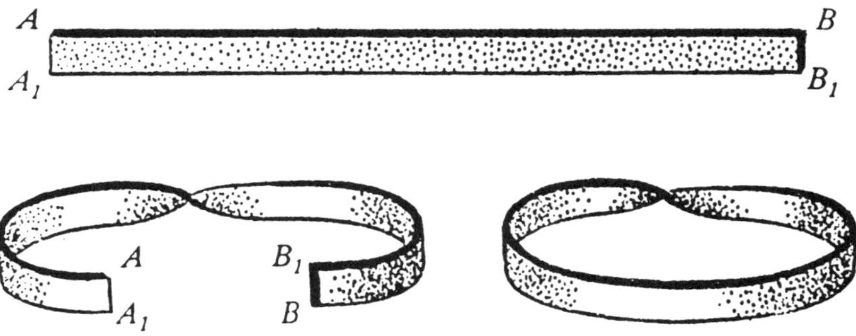

Figure 6.2

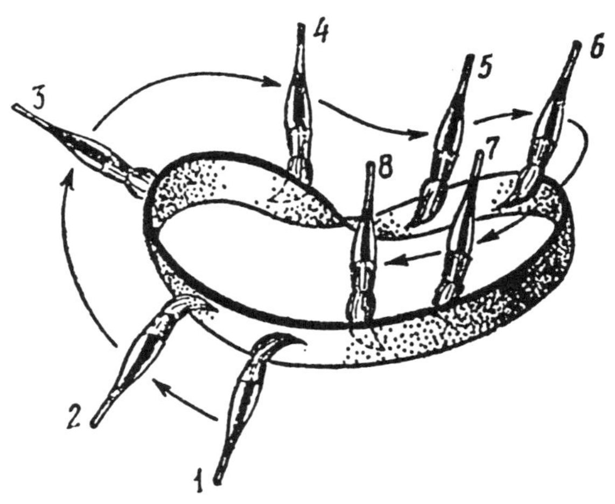

Figure 6.3

Exercise 6.4. We know that the 4×9 rectangle cannot be tiled by L-tetrominoes. Prove that the Möbius band $M(4 \times 9)$ can be tiled by L-tetrominoes.

Exercise 6.5. Let P be a polyomino. It is known that exactly one of the two Möbius bands $M(m \times n)$, $M(n \times m)$ can be tiled by copies of P. What can you conclude about the tileability of the plane $m \times n$ rectangle?

So far in this section we compared the tileability of a $m \times n$ rectangle with the tileability of the corresponding cylinder $C(m \times n)$, torus $T(m \times n)$, and Möbius band $M(m \times n)$. Let us now look at the not any less exciting *problem of comparing the tileability of the above surfaces with the tileability of the plane and an infinite strip.*

Exercise 6.6. Prove that if there is a rectangle R tileable by, say, tiles of shape T, then so is the plane.

Is the converse of the statement of Exercise 6.6 true?

Exercise 6.7. The plane is tileable by tiles of shape T. Is it true that then there exists a rectangle R tileable by tiles of shape T?

Let $C(m \times n)$ be a cylinder tiled by copies of a polyominal tile T. We take a "cut" from the top of the cylinder to its bottom, by a vertical line that does not lie on the edges of the grid. Then we take the union of the tiles that were cut, and consider the right boundary S of the union of this set of tiles. See Figure 6.4 below, in which the "cut" is indicated by a dotted line, the tiles that are cut are marked by black dots, and the right boundary, which is the "*step-line*" S to be used, is heavily drawn.

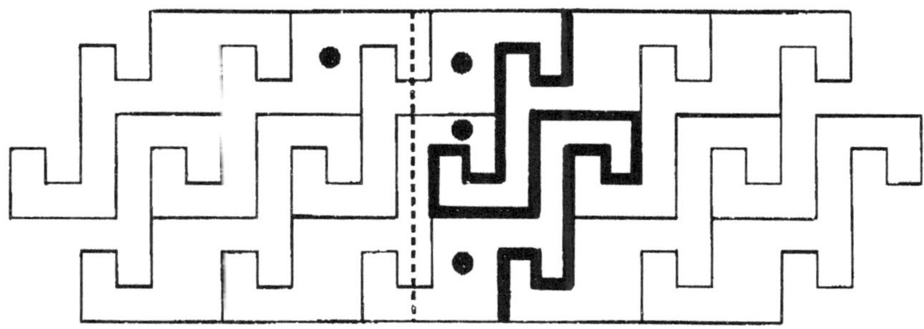

Figure 6.4

Now we cut the cylinder along S, and then open up and flatten its surface. As a result we get a plane figure F with parallel top and bottom boundaries and "parallel" left and right

54

"step-line" boundaries. Now it is clear how to formalize these notions:

Step-line is the one that can be traced continuously along the lines of a checkered grid with a pencil without self-intersections.

We will call two step-lines *parallel* if one of them is an image of the other under a translation.

The opposite sides of the figure F are parallel. Let t be a translation that maps the left boundary of F onto its right boundary. Then by repeatedly applying the translation t and the *inverse* translation t^{-1} (i.e., a move in the opposite direction through the same distance) to F and its consequent images, we will tile an infinite strip S with copies of F (Figure 6.5).

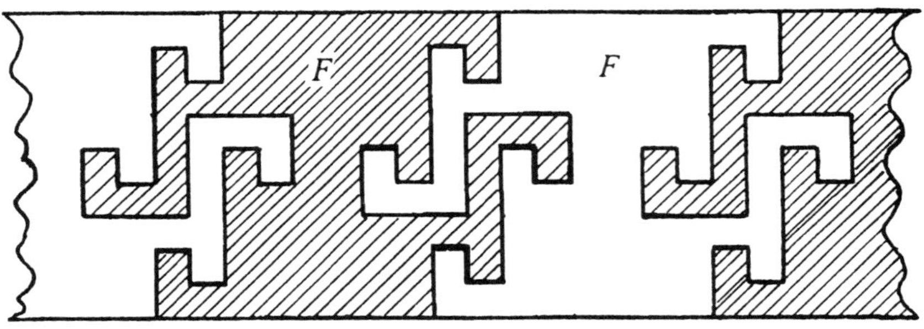

Figure 6.5

Since the figure F itself is tiled by tiles of shape T, we get the tiling of the infinite strip S by tiles T. This type of tiling of an infinite strip by tiles T is called *periodic*: it is obtained by repeating applications of a translation t (and t^{-1}) to a tiling of a bounded figure F.

We proved that *every tiling of a cylinder leads to a periodic tiling of an infinite strip*. Now you are ready for two very exciting questions that were contributed by the famous geometer, Branko Grünbaum, a coauthor of the great book, "Tiling and Patterns" ([GS]).

Exercise 6.8. (B. Grünbaum) Is there a tiling of an infinite strip by copies of a tile T that is not periodic?

Exercise 6.9. (B. Grünbaum) Prove that every tiling of an infinite strip by copies of a polyominal tile T contains a bounded part that can be used to tile a cylinder. (Hint: please read the first paragraph of Section 8).

Let $T(m \times n)$ be a torus tiled by copies of a polyominal tile T. We can picture the torus $T(m \times n)$ as the $m \times n$ rectangle R but we must remember that the opposite sides of R are glued together. Now we can cut the torus by two step-lines along some boundaries of adjacent tiles, so that one step-line goes from the top to the bottom edge of R, and the other from the left to the right edge of R (Figure 6.6).

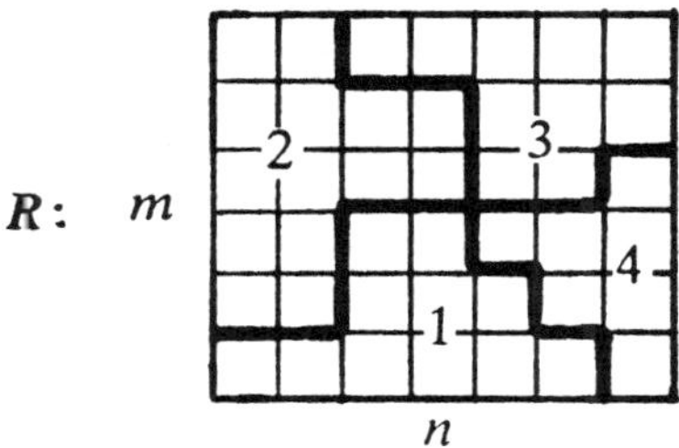

Figure 6.6

Now we combine the four parts of R from Figure 6.6 together to form a figure F (see Figure 6.7).

Let t_1 be the translation to the right through n and t_2 the translation upward through m. The figure F has two useful properties:

i) By repeatedly applying the translations t_1 and t_2 and their inverses t_1^{-1} and t_2^{-1} to F and its consequent images, we will tile the entire plane by copies of F.

ii) F can be tiled by copies of tile T (that originally tiled the given torus $T(m,n)$).

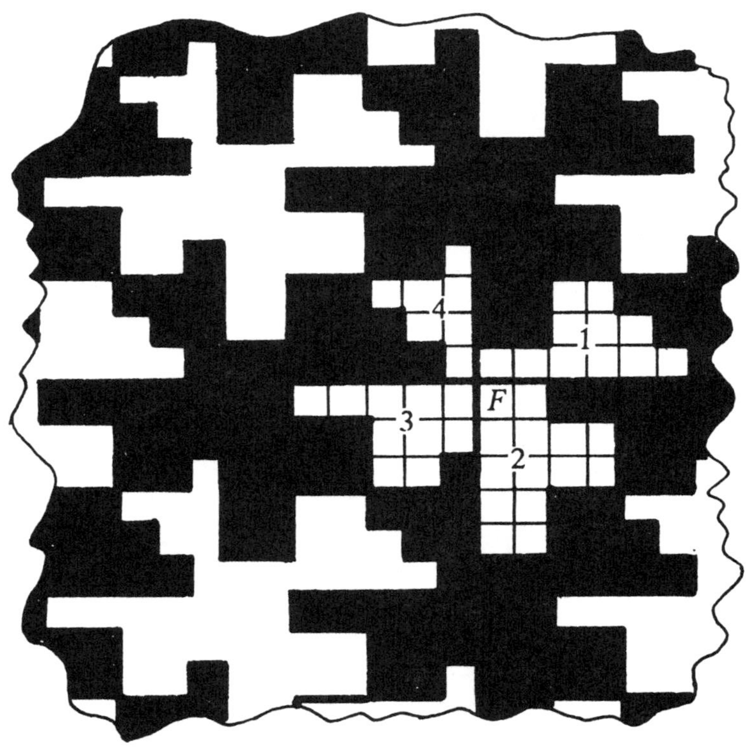

Figure 6.7

We leave the pleasure of proving the properties i) and ii) for the readers. These two properties give birth to the tiling of the plane by copies of tile T. This type of tiling of the plane is called **periodic**: it is obtained by repeating applications of two translations t_1 and t_2 (and their inverses) to a tiling of a bounded figure F.

We proved here that *every tiling of a torus leads to a periodic tiling of the plane.*

We do not know the answer to the following question contributed by Branko Grünbaum. The only consolation is that apparently nobody does!

Open Problem 6.1. *Is it true that every tiling of the plane by copies of a polyominal tile contains a bounded part that can be used to tile a torus?*

We are offering \$50.00 for the first solution of this problem. (The authors will be the judges of what consitutes a solution.)

Tiling a rectangular board can be expanded in a different way: by adding a dimension (or a few!) to our tiling games. In three dimensions we call it *packing* a parallelepiped (Figure 6.8).

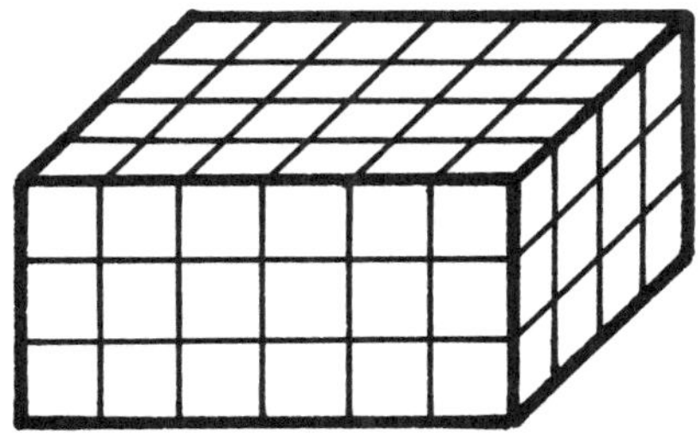

Figure 6.8

Example 6.1. *Find all positive integers m, n, and k such that the parallelepiped $m \times n \times k$ can be packed by linear 3-dimensional p-ominoes (i.e., $1 \times 1 \times p$ bricks, see Figure 6.9).*

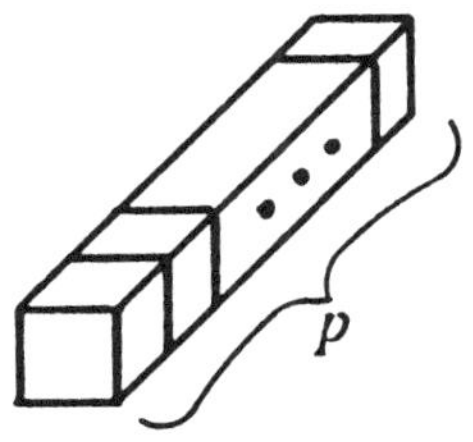

Figure 6.9

Solution. Assume that a $m \times n \times k$ parallelepiped P can be packed by linear 3-dimensional p-ominoes, but neither m nor n are divisible by p. We will prove that then k is divisible by p.

Let us color all *mnk* unit squares of P in p colors by coloring that is meaningful to call *the direct product of diagonal and column colorings* or in short $D \times C$ coloring (see Figures 4.1, 4.2, and 4.3). In other words, the lower horizontal $m \times n$ layer of unit cubes we color diagonally in p colors with cyclic permutation of colors (see Figures 4.2 and 4.1). The parallelepiped can be viewed as a union of mn vertical columns "growing" up each from a foundation, i.e., from a unit cube of the lower level. We color all the unit cubes of a column in the same color as its foundation; thus we get mn one-color columns.

You undoubtedly noticed that in $D \times C$ coloring a linear p-omino placed horizontally covers one unit cube of each of the p colors, and each vertical linear p-omino covers p cubes of the same color. Therefore, for any tileable parallelepiped the numbers S_i of unit cubes of color i ($i = 1, 2, \ldots p$) are congruent modulo p:

$$S_1 \equiv S_2 \equiv \ldots S_p \ (\mathrm{mod}\ p) \qquad (*)$$

The given parallelepiped P can be packed by linear p-ominoes by assumption; therefore, it satisfies the condition $(*)$.

Just as in the first proof of Theorem 4.1, we take advantage of the fact that there are non-negative integers q and r_1 such that

$$0 \leq r_1 \leq \frac{p}{2} \text{ and}$$

$$m = pq + r_1$$

or

$$m = pq - r_1$$

Accordingly, we have to consider two cases:

Case 1. $m = pq + r_1$. We cut the parallelepiped P into two parallelepipeds: $pq \times n \times k$ and $r_1 \times n \times k$. Since the $pq \times n \times k$ parallelepiped can be packed by linear p-ominoes, it satisfies condition $(*)$. Since congruence $(*)$ is true for the parallelepiped P, it is also true for the $r_1 \times n \times k$ parallelepiped P_1.

Case 2. $m = pq - r_1$. Let us attach an $r_1 \times n \times k$ parallelepiped P_1 to the given parallelepiped P to obtain the $pq \times n \times$

k parallelepiped P' and extend the coloring of P to the $D \times C$ coloring of P'. Since P' can be packed by linear p-ominoes, P' satisfies condition (*). Thus, both P' and P satisfy condition (*); therefore, P_1 satisfies condition (*).

Thus, in both cases we got an $r_1 \times n \times k$ parallelepiped P_1 with the $D \times C$ coloring, such that $0 \leq r_1 \leq \frac{p}{2}$ and P_1 satisfies condition (*). Now we can turn P_1 about a vertical axis through a 90° angle, i.e., consider the $n \times r_1 \times k$ parallelepiped, and apply to it *all* of the above reasoning. As a result, we will get an $r_2 \times r_1 \times k$ parallelepiped P_2 with $0 \leq r_2 \leq \frac{p}{2}$, that satisfies condition (*), i.e.,

$$S_1 \equiv S_2 \equiv \ldots S_p \ (mod \ p) \tag{*}$$

On the other hand, the number of one-color unit squares in the $r_2 \times r_1$ rectangle is equal to $r_2 + r_1 - 1$, and

$$r_2 + r_1 - 1 \leq \frac{p}{2} + \frac{p}{2} - 1 < p$$

Therefore, the p-th color is not present in the parallelepiped P_2 at all. I.e.,

$$S_p = 0$$

and we have exactly one column of the first color:

$$S_1 = k$$

These two equalities together with the congruence (*) show that $k \equiv 0 \ (\mathrm{mod} \ p)$, i.e., k is divisible by p.

Thus a parallelepiped $m \times n \times k$ can be packed by linear 3-dimensional p-ominoes if and only if at least one of the numbers m, n, k is divisible by p.

Exercise 6.10. Find all positive integers m, n, and k such that the $m \times n \times k$ parallelepiped can be packed by 3-dimensional L-trominoes (Figure 6.10).

Exercise 6.11. Let k and n be positive integers, $k < n$. Prove that no matter which $kn^2 + 1$ unit cubes of the $n \times n \times n$ cube C are colored red, you can always choose $k + 1$ red cubes such that no two of them lie in the same layer of C. (There are

60

three different types of layers in C that are parallel to various faces of C.)

Figure 6.10

In fact, Exercise 6.11 allows an n-dimensional generalization that was noticed by A. Soifer and S. Slobodnik in 1973 (see [SS], [S1]), and will appear with a proof and related problems in [S4]. For $n = 10$ this generalization can be nicely formulated in the language of numbers in the decimal system:

Given $r \cdot 10^{n-1} + 1$ distinct n-digit numbers, $0 < r < 9$. Prove that you can choose $r + 1$ numbers out of them, such that any two of the $r + 1$ numbers in any decimal location have distinct digits.

Solutions of Exercises

Figure 6.11

6.1. As we know from Section 2, the $n \times n$ square can be tiled by L-trominoes if and only if n is divisible by 3 and not equal to 3. On the torus this exception is not necessary: the torus $T(3 \times 3)$ can be tiled (Figure 6.11).

Indeed, when we form the torus from the 3×3 square, the three black squares get together and form an L-tromino!

6.2. Yes. The cylinder $C(4 \times 2)$ can be tiled by T-tetrominoes (Figure 6.12). The cylinder $C(2 \times 4)$ can not be tiled.

Figure 6.12

Indeed, when we glue the top and bottom of the 4×2 rectangle together, the black squares come together and form a T-tetromino (Figure 6.12). The cylinder $C(2 \times 4)$ cannot be tiled because two squares in the upper row of the 4×2 rectangle must be covered by two distinct T-tetrominoes T_1 and T_2, which in turn makes T_1 and T_2 overlap.

6.3. The cylinder $C(m \times n)$ can be tiled by linear k-ominoes if and only if at least one of the numbers m, n is divisible by k (in other words, we do not gain anything compared to tiling a plane $m \times n$ rectangle).

To prove this, we color rings of the $C(m \times n)$ cylinder in k colors with cyclic permutation of colors (Figure 6.13), and then proceed exactly like in the second solution of Problem 4.1 in Section 4.

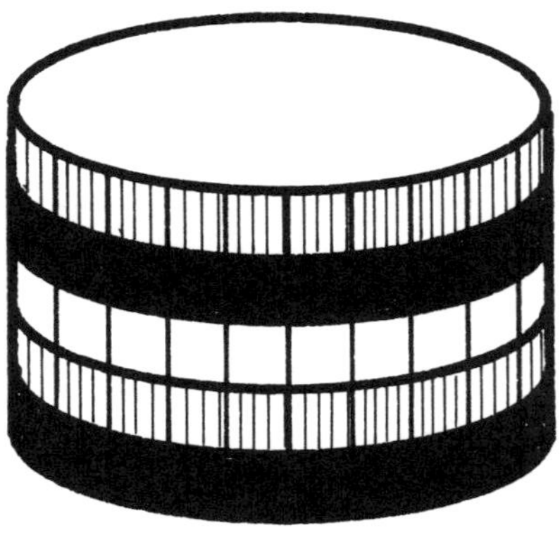

Figure 6.13

6.4. A solution is shown in Figure 6.14.

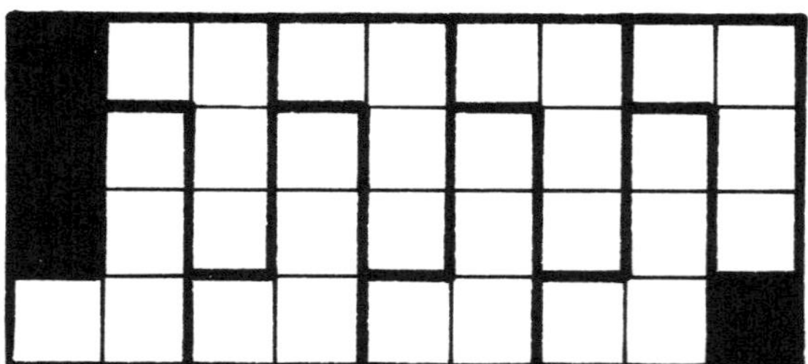

Figure 6.14

When you glue the Möbius band out of this rectangle, the two black regions join together, creating an L-tetromino.

6.5. Assume that the $m \times n$ rectangle can be tiled by copies of a polyomino P. Then both Möbius bands $M(m \times n)$ and $M(n \times m)$ will be tileable by copies of P, whereas we are given that exactly one of them is tileable. Therefore, the $m \times n$ rectangle cannot be tiled by copies of P.

6.6. Copies of R easily tile the plane (Figure 6.15). Tiling each copy of R with tiles of shape T will thus generate the tiling of the plane by tiles of shape T.

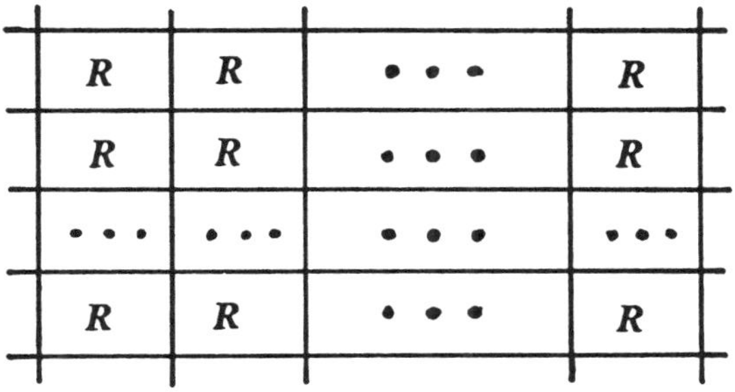

R	R	$\bullet\ \ \bullet\ \ \bullet$	R
R	R	$\bullet\ \ \bullet\ \ \bullet$	R
$\bullet\ \bullet\ \bullet$	$\bullet\ \bullet\ \bullet$	$\bullet\ \ \bullet\ \ \bullet$	$\bullet\ \bullet\ \bullet$
R	R	$\bullet\ \ \bullet\ \ \bullet$	R

Figure 6.15

6.7. No! We can tile an infinite strip by Z-tetrominoes (Figure 6.16).

Figure 6.16

Such strips easily tile the plane. On the other hand, in Exercise 3.2 we proved that there is no rectangle tileable by Z-tetrominoes.

6.8. There is! We can split the 4-wide infinite strip into 4×2 rectangles, and put in each of them one digit "0" or "1" alternating them as follows: ... 10000100010010100100001 ... (Figure 6.17).

... | 1 | 0 | 0 | 0 | 0 | 1 | 0 | 0 | 0 | 1 | 0 | 0 | 1 | 0 | 1 | 0 | 0 | 1 | 0 | 0 | 0 | 1 | 0 | 0 | 0 | 0 | 1 | ...

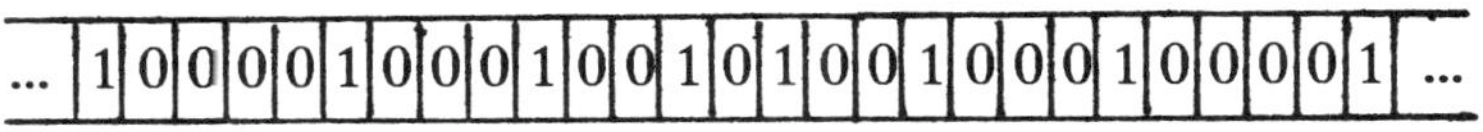

Figure 6.17

64

Now we tile all the 4×2 rectangles with "0" as in Figure 6.18a. All the 4×2 rectangles with "1" we tile as shown in Figure 6.18b. As a result we get a non-periodic tiling of the infinite strip by L-tetromino (can you prove it?).

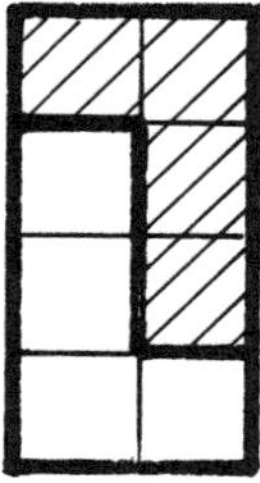

Figure 6.18a

Figure 6.18b

6.9. Please see Exercise 8.5.

6.10. Let $m \geq n \geq k$. If $k = 1$, then the solution for m and n is exactly the same as in Theorems 3.1 and 3.2 of Section 3.

Let $k \geq 2$. You can show (do) that the $3 \times 2 \times 2$, $3 \times 3 \times 2$, and $3 \times 3 \times 3$ parallelepipeds can be packed by 3-dimensional L-trominoes. This implies (can you show how?) that the $m \times n \times k$ parallelepiped can be packed by 3-dimensional L-trominoes if and only if the product mnk is divisible by 3.

6.11. This problem is a 3-dimensional analog of Exercise 7.6 discussed in the next section.

CHAPTER 2

PROOFS OF EXISTENCE

7. THE PIGEONHOLE PRINCIPLE IN GEOMETRY

We'll start by offering the reader the opportunity to prove the **Pigeonhole Principle** itself. It is also known as the *Dirichlet Principle,* after the famous mathematician Peter Gustav Lejeune Dirichlet (1805-1859).

Exercise 7.1. (*Pigeonhole Principle*). If $kn + 1$ pigeons (k and n are positive integers) sit on n pigeonholes, then at least one of the holes has at least $k + 1$ pigeons on it.

Please note that the principle guarantees the *existence* of a hole with lots of pigeons on it, but as it often happens in mathematics, it gives us no way to find this hole.

Example 7.1. *Prove that among any five points located inside or on the boundary of a unit square there are two points not more than $\frac{1}{\sqrt{2}}$ apart.*

Solution. As you can see *pigeons and pigeonholes are not given to us. We have to invent them if we wish to use the pigeonhole principle.*

Let us partition the given unit square into four quarter squares (Figure 7.1) These quarter squares will be our pigeonholes, and, of course, the five given points will serve as the pigeons.

Since $5 = 1 \cdot 4 + 1$, in view of the pigeonhole principle, there is a pigeonhole that contains at least two pigeons. In other words, there is a quarter square that contains at least two given points A,B. The distance $|AB|$ is, of course, no greater than the diagonal $\frac{1}{\sqrt{2}}$ of a quarter square.

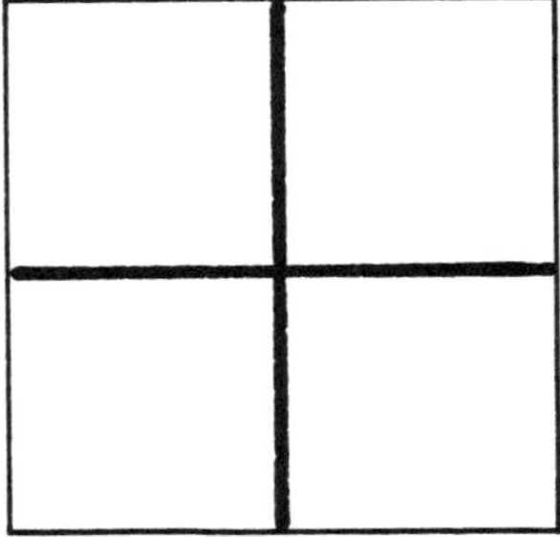

Figure 7.1

Example 7.2. *Prove that among any six points located in a 3 × 4 rectangle there are at least two points not more than $\sqrt{5}$ apart.*

Solution. Sometimes we need to allow the pigeonholes to be different in size and shape. In order to solve this problem, let us partition the 3 × 4 rectangle into five polygons that will be our pigeonholes (Figure 7.2).

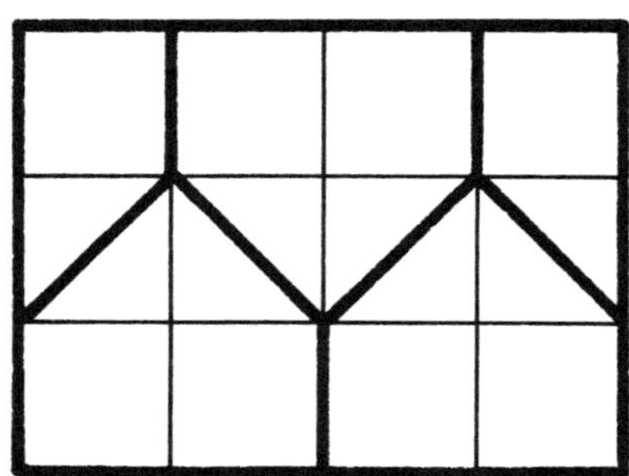

Figure 7.2

Six pigeons (the given points) sit on five pigeonholes, therefore at least two pigeons sit on the same pigeonhole. As you can easily compute, the maximal distance between any two "pigeons" inside the same pigeonhole is $\sqrt{5}$.

Exercise 7.2. Prove that no matter how a plane is colored in two colors, it contains two points of the same color exactly one mile apart.

Exercise 7.3. (A. Soifer [S2]) Prove that among any nine points inside or on the boundary of a triangle of area 1, there are three points that form a triangle of area not exceeding $\frac{1}{4}$.

We can improve the result of the previous exercise:

Exercise 7.4. (A. Soifer [S2]) Prove that among any seven points inside or on the boundary of a triangle of area 1, there are three points that form a triangle of area not exceeding $\frac{1}{4}$.

The result of Exercise 7.4 can be improved too: see this in the book [S2].

Exercise 7.5. All vertices of a convex pentagon lie on the intersections of a grid. Prove that the pentagon (i.e., interior plus boundary) contains at least one more intersection of the grid.

Exercise 7.6. (A. Soifer and S. Slobodnik [SS], [S1]) Forty-one rooks are placed on a 10×10 chessboard. Prove that you can choose five of them that do not attack each other. (We say that one rook *attacks* another if they are in the same row or column of the chessboard.)

Solutions of Exercises

7.1. Assume that there are no holes that contain at least $k + 1$ pigeons. Then

$$\text{the 1st hole contains} \leq k \text{ pigeons}$$
$$\text{the 2nd hole contains} \leq k \text{ pigeons}$$
$$\vdots$$
$$\underline{\text{the } n\text{th hole contains} < k \text{ pigeons}}$$
$$\text{the total number of pigeons} \leq k \times n$$

This contradicts the given fact that there are $kn + 1$ pigeons. Therefore, there is a hole that contains at least $k + 1$ pigeons.

7.2. Look at the vertices of an equilateral triangle with side one mile on the colored plane. Since its three vertices (pigeons) are painted in two colors (pigeonholes), we can choose two vertices painted in the same color.

This problem is a starting point of a celebrated open problem. The same statement (i.e., the *existence of two points of the same color 1 unit apart*) can be proven even if the plane is colored in three colors (try to prove it). It is also known that there is a coloring of the plane in seven colors that prevents the existence of two points of the same color 1 unit apart (try to show this too!).

The question is still open for four, five, and six colors after many years and numerous attempts to solve this problem.

7.3. Midlines partition the given triangle into four congruent triangles of area $\frac{1}{4}$ (Figure 7.3).

These congruent triangles are our pigeonholes, and the given points are our pigeons. Now nine pigeons are sitting in four pigeonholes. Since $9 = 2 \times 4 + 1$, there is at least one pigeonhole containing at least three pigeons.

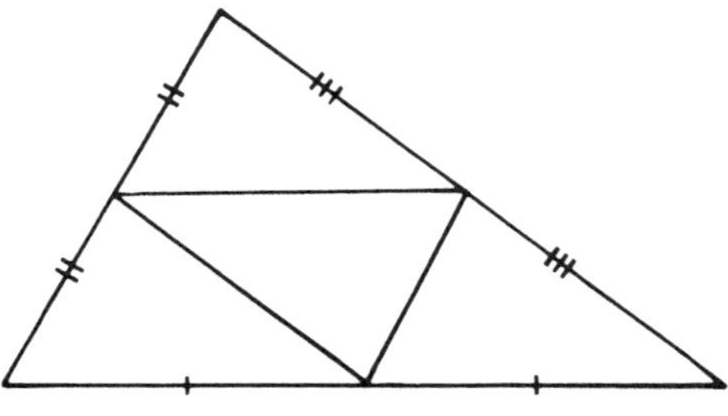

Figure 7.3

7.4. Since $7 = 2 \times 3 + 1$, *it would be nice to have three pigeonholes*--then at least one of them would have at least three pigeons! Okay, let's draw only two midlines in the given triangle (Figure 7.4).

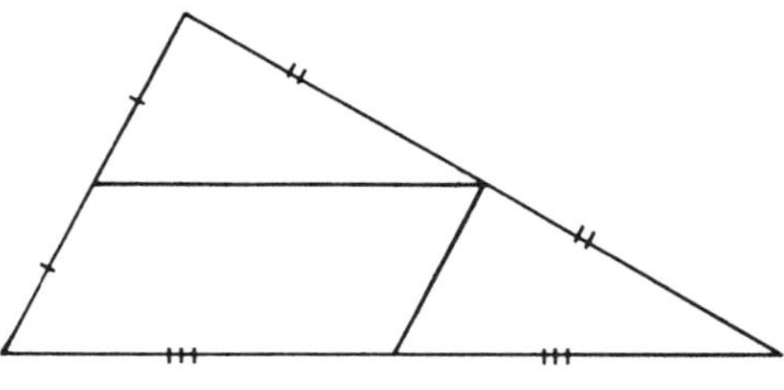

Figure 7.4

We get three pigeonholes. At least one of them contains at least three pigeons. If one of the triangles contains three given points, we're done.

If the parallelogram contains three given points, then all we have left to prove is a simple lemma: *The maximum area of a triangle inscribed in a parallelogram of area 1/2 is equal to 1/4.* We leave the proof of this lemma to the reader.

7.5. Let's introduce the coordinate system on the grid (Figure 7.5).

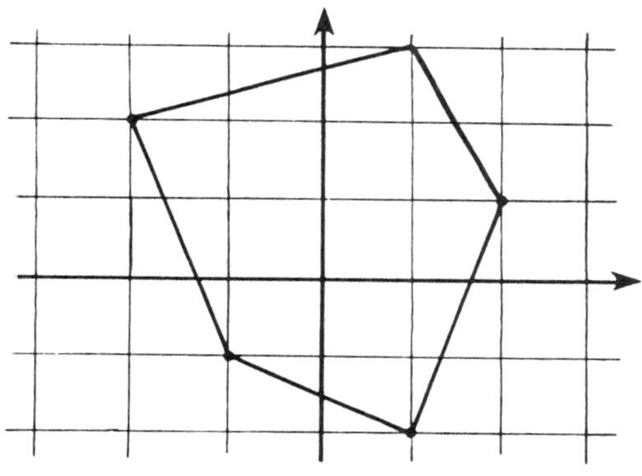

Figure 7.5

To each vertex M of the pentagon with coordinates (x,y) we assign the ordered pair of the remainders upon division of coordinates x,y by 2. There are only four possible outcomes of this operation: $(0,0)$, $(0,1)$, $(1,0)$, and $(1,1)$. These are our pigeonholes. Since we have five pigeons (the vertices of the pentagon), there are two vertices, M_1 and M_2, that give the same pair of remainders. In order to complete the proof, all there is left to notice is that the midpoint of the segment M_1M_2 has integral coordinates and lies on the interior or boundary of the pentagon.

7.6. Let's make a cylinder out of the chessboard by gluing together two opposite sides of the board and color the cylinder diagonally in 10 colors (Figure 7.6).

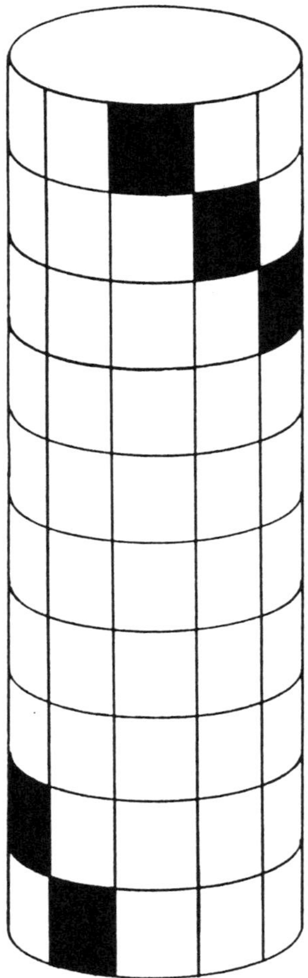

One out of the ten one-color diagonals is shown in black.

Figure 7.6

Now we have $41 = 4 \times 10 + 1$ pigeons (rooks) in 10 pigeonholes (one-color diagonals). Therefore, there is at least one hole containing at least 5 pigeons. But the 5 rooks located on the same one-color diagonal do not attack each other!

8. AN INFINITE FLOCK OF PIGEONS

What would happen if an infinite flock of pigeons were to land on a finite number of pigeonholes? The answer is clear: at least one of the holes would contain an infinite subflock! We will call this simple argument the *Infinite Pigeonhole Principle*. It will allow us to solve a number of important problems.

Let M be an infinite *bounded subset* of the real line R. The word "*bounded*" means that there exists a positive number m, such that $|x| < m$ for every x that belongs to M. A point p of R is said to be a *limit point* of the set M, if for every $\epsilon > 0$ the segment $[p - \epsilon, p + \epsilon]$ contains infinitely many points of M.

Now we are ready for a classical result of mathematics.

Example 8.1 *(Bolzano-Weierstrass Theorem)*. Every infinite bounded subset M of the real line R has at least one limit point.

Proof. Imagine points of M as pigeons. So, M is an infinite flock of pigeons. And what do we take as pigeonholes? Since M is bounded, there is a positive integer m such that $|x| < m$ for every x from M. Now we take the segments

$$[-m, -m + 1], [-m + 1, -m + 2], ..,$$
$$[-1, 0], [0, 1], ..., [m - 1, m] \qquad (*)$$

as pigeonholes. Then due to the infinite pigeonhole principle, at least one of them, say, $[1,2]$, contains infinitely many pigeons. Thus, we have infinitely many pigeons that in decimal form are equal to $1. a_1 a_2 a_3 ...$ where $a_1, a_2, a_3, ...$ are their digits.

The next step is to consider ten segments.

$$[1.0, 1.1], [1.1, 1.2], ..., [1.8, 1.9], [1.9, 2.0] \qquad (**)$$

as new pigeonholes. Again we conclude that one of them, say, $[1.4, 1.5]$, contains infinitely many pigeons. Thus, we have infinitely many pigeons of the form $1, 4 a_2 a_3 ...$

Now from ten segments

$$[1.40,\ 1.41],\ [1.41,\ 1.42],\ ...,\ [1.49,\ 1.50] \qquad (***)$$

we choose one, say, $[1.43,\ 1.44]$, that contains infinitely many pigeons, and so on.

We obtain a real number $x = 1.43...$ We can easily show that x is a limit point of the set M. Indeed, given a positive number ϵ, we choose an integer m, such that $\frac{1}{10^m} < \epsilon$. Let, further, $x_m = 1.43\,a_3 a_4... a_m$ be a *finite* decimal fraction that we obtain from x by removing all decimal digits of x except the first m digits after the decimal point. Then the segment $[x_m,\ x_m +\frac{1}{10^m}]$ contains infinitely many pigeons and is contained in the segment $[x - \epsilon,\ x + \epsilon]$. This completes the proof.

Let M be a non-empty bounded subset of the real line R. A point a of R is said to be the *exact upper bound* of the set M if it possesses both following properties:

i)   there is no point x in M, such that $x > a$;

ii)   for every positive number ϵ, the segment $[a - \epsilon,\ a]$ contains at least one point of the set M.

The *exact lower bound* of M is defined similarly.

It is clear that if the exact upper bound a of M belongs to M, then a is the *maximal* point of M. If not, then a is a limit point of M. The following is a classical theorem of analysis.

Example 8.2. *Every non-empty bounded subset of R has the exact upper bound (and the exact lower bound).*

Proof. Consider again the pigeonholes (*). Then at least one of them contains a point of M. We take the *utmost right* pigeonhole that contains a point of M. Let it be the pigeonhole $[1,2]$. Then we consider ten pigeonholes (**) and take the utmost right pigeonhole that contains a point of M. Let $[1.4,\ 1.5]$ be this pigeonhole. Then we consider the pigeonholes (***) and so on. As a result, we obtain a real number $a = 1.43....$ It can be easily shown (do) that a is the exact upper bound of M.

Let now M be an infinite **bounded point set** in the plane R^2. The word "*bounded*" means that there exists a square (or a disk) that contains M. Here and everywhere in this book by "*disk*" we mean "closed circular disk," i.e., a circle together with all the points inside it. A point p of R^2 is said to be a ***limit point*** of M, if each disk with center p contains infinitely many points of M.

Example 8.3. *Every infinite bounded subset M of the plane R^2 has at least one limit point.*

Proof. Indeed, let m be a positive integer, such that for every point $q = (x,y)$ of the set M, the coordinates x,y satisfy the inequalities $|x| < m$, $|y| < m$. We divide the square with the vertices $(\pm m, \pm m)$ into unit squares (Figure 8.1). These unit squares are the pigeonholes, and all points of M are the pigeons. By the infinite pigeonhole principle at least one pigeonhole P

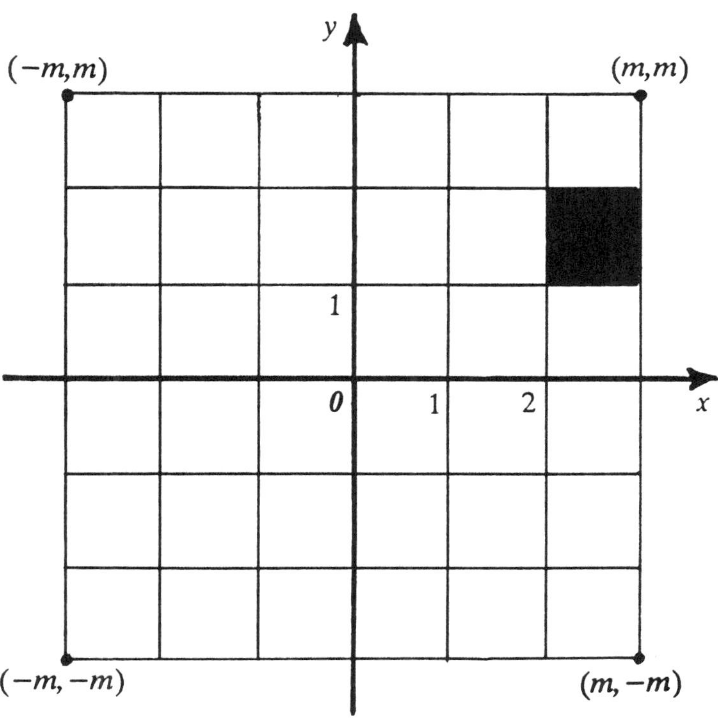

Figure 8.1

contains infinitely many pigeons. Let, for example, the point (2,1) be the left bottom corner of P. Then each pigeon in this pigeonhole has the form $q = (x,y)$ where $x = 2.a_1 a_2 a_3...$ and $y = 1.b_1 b_2 b_3...$.

We now divide the pigeonhole P into 100 new pigeonholes, i.e., 100 small squares with the length of sides $\frac{1}{10}$. Again, there exists a pigeonhole P_1 that contains infinitely many pigeons. Let, for example, the point (2.7, 1.4) be the left bottom corner of P_1. Then each pigeon in P_1 has the form $q = (x,y)$ where $x = 2.7a_2 a_3 ...$ and $y = 1.4b_2 b_3 ....$ Now we divide P_1 into 100 pigeonholes that are small squares with the length of sides $\frac{1}{100}$, and chose a pigeonhole P_2 with the left bottom corner, say (2.76, 1.43), that contains infinitely many pigeons, and so on.

We obtain a pair of real numbers $q = (x,y)$ where $x = 2.76...$, $y = 1.43 ....$ It can be easily shown (do) that q is a limit point of M. We are done.

A similar assertion is true for the space R^3 and even for n-dimensional space R^n, where n is an arbitrary positive integer.

The examples we discussed above allow us to introduce the notion of a compact set which is important for the last chapter of this book. A set M in the plane R^2 (similarly in R^n for arbitrary positive integer n) is said to be *closed* if it contains all its limit points. A closed bounded set is said to be *compact*.

The following proposition, known as the **Weierstrass Theorem**, indicates an important property of continuous functions on compact sets:

Each continuous function f defined on a compact set M (Figure 8.2) is bounded and reaches at at least one point a of M its maximal value (a similar statement is true for the minimal value).

We do not give here the exact definition of continuity (that is, "absence of breaks"), assuming that this notion has a clear visual meaning. Perhaps in one of the following issues of our *Etudes* we will return to geometrical ideas in analysis and, in particular, to the idea of continuity.

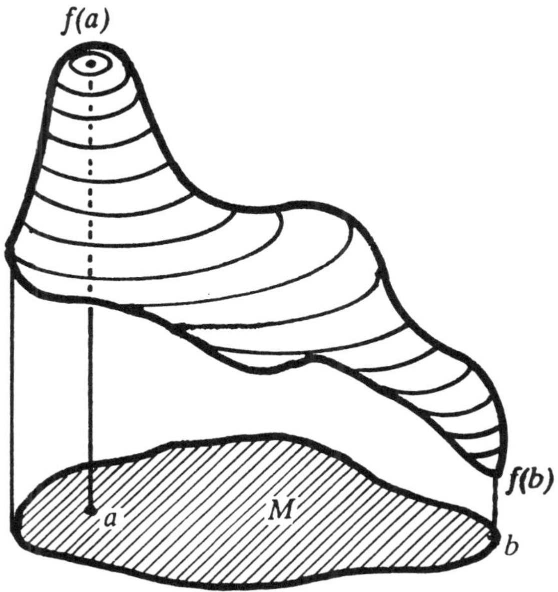

Figure 8.2

The following proposition, known as the **Intermediate Value Theorem**, expresses another important property of continuous functions:

Let M be a connected set (that is, M consists of "one piece") and f be a continuous function on M. If now a and b are two points of M and y is a number, such that $f(a) < y < f(b)$, then there exists a point c of M for which $f(c) = y$ (Figure 8.3).

Exercise 8.1. A compact figure F of area S and a line L are given in the plane. Prove that there exists a line parallel to L that divides F into two parts each of area $\frac{1}{2}S$ (Figure 8.4).

Exercise 8.2. A compact figure F of area S and a point p are given in the plane. Prove that there exists a line through p that divides F into two parts each of area $\frac{1}{2}S$.

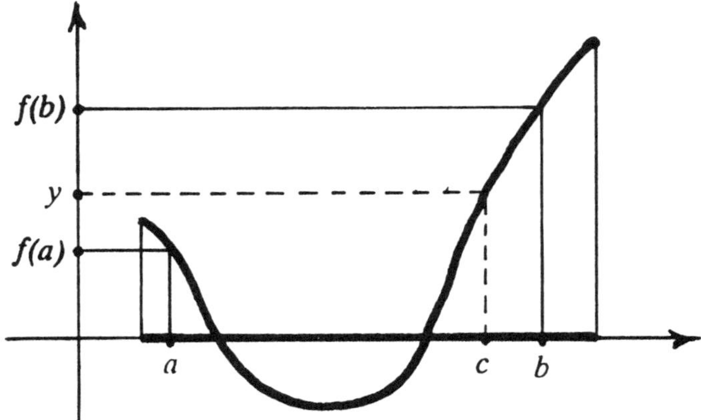

Figure 8.3

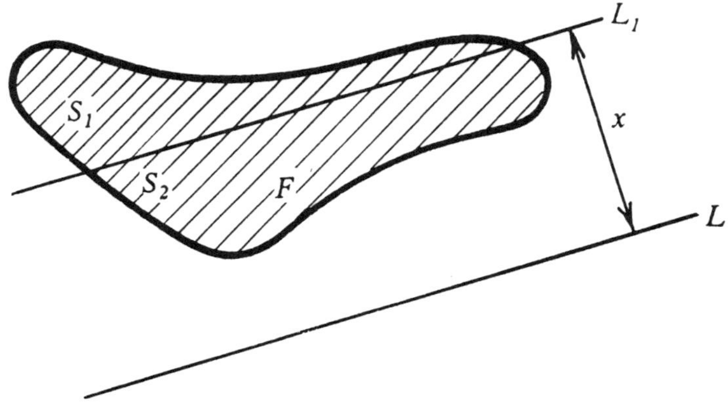

Figure 8.4

Exercise 8.3. A Russian mathematician,Pavel Uryson,noticed that for every compact figure F in the plane there exists a square circumscribed about F. Can you prove that?

The above exercises and similar problems can be found in the book [S2] and the article [T].

Let M_1, M_2, ... be arbitrary figures in the plane (or in R^n). The *intersection* $M_1 \cap M_2 \cap$... of these figures is the set of all points that belong to *all* the figures M_1, M_2, ... If there is at least one common point of all the figures M_1, M_2, ..., then we say that the intersection of these figures is *non-empty*.

Example 8.4. *Let us consider a decreasing sequence M_1, M_2, ... of compact figures in the plane (or in R^n), that is, M_j is contained in M_i if $j > i$. Prove that the intersection of all figures M_i (that is, the set of points that belong to all the figures) is a non-empty compact figure.*

Proof. Indeed, the intersection of closed sets is always closed. The intersection of bounded figures is bounded. So, it remains to prove that the intersection I of a decreasing sequence of non-empty compact figures M_i is non-empty. Assume the opposite, i.e., I is empty. We take a point a_1 of the figure M_1. Then a_1 does not belong to I (since I is empty). So there exists an index i, such that a_1 does not belong to M_i. Without loss of generality we can assume that a_1 does not belong to M_2 (if necessary we just throw away the figures M_2, ..., M_{i-1} and rename M_i by M_2. Now we choose a point a_2 of M_2. Then a_1 is distinct from a_2. Again, a_2 does not belong to I and, consequently, a_2 does not belong to a figure M_j for some index j. Without loss of generality we can assume that a_2 does not belong to M_3. Continuing, we obtain a sequence a_1, a_2, ... of distinct points, such that a_i belongs to M_i, but does not belong to M_{i+1}, M_{i+2}, ... Moreover, the set $\{a_1, a_2, a_3, ...\}$ is bounded (it is contained in M_1).

According to the Bolzano-Weierstrass Theorem there is a limit point b of the set $\{a_1, a_2, a_3, ...\}$. Since all the points a_1, a_2 a_3, ... except a_1, a_2, ..., a_{i-1} belong to M_i, b also belongs to M_i (we recall that M_i is closed, that is, it contains all its limit points). Thus, each figure M_i ($i = 1, 2, ...$) contains b, contradicting the assumption that the intersection of all these figures is empty.

Let us note that the theorem we have just proved is not true for non-compact figures (even closed). For example, let us denote by M_i the half-plane determined in a coordinate system (x,y) by the inequality $x \geq i$ (where $i = 1, 2 ...$). We obtain a decreasing sequence of closed non-compact figures M_1, M_2, ..., such that their intersection is empty (see Figure 18.14). We are done.

Now let us introduce a new notion. Let M be a figure in the plane, and ϵ a positive number. A point x is said to be *ϵ-close* to the figure M, if there exists a point y of M, such that the distance $|xy|$ between these points does not exceed ϵ. The set $E_\epsilon(M)$ of of all points x that are ε-close to M is called the *ε-extension* of the figure M (Figure 8.5). In other words, the ε-extension of M is the union of all disks of radius ε with centers in the points of the figure M. It can be easily shown that if M is compact, then its ε-extension is also compact.

The German mathematician F. Hausdorff introduced a notion of *distance* between two compact figures. Here is his definition. Let M_1 and M_2 be two distinct compact figures in the plane. The *distance* $d(M_1, M_2)$ between M_1 and M_2 is the least positive number ε, such that M_1 is contained in the ε-extension of M_2 and M_2 is contained in the ε-extension of M_1.

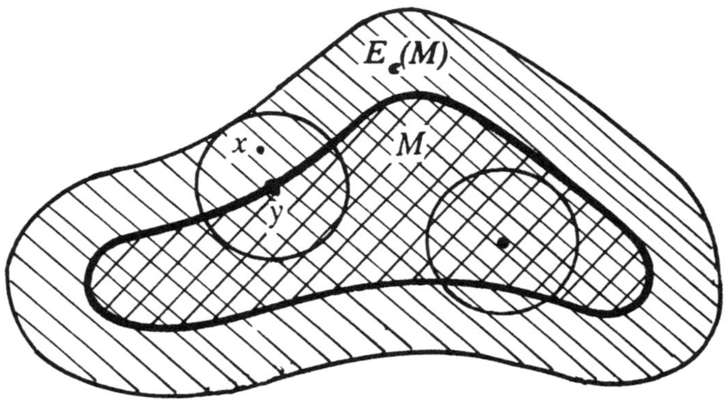

Figure 8.5

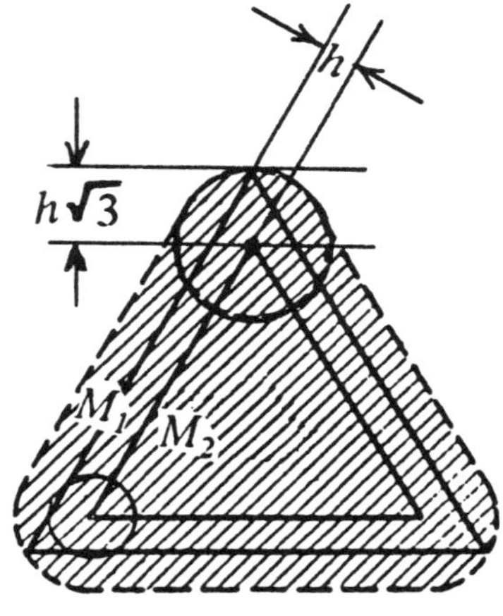

Figure 8.6

For example, if M_1 and M_2 are two equilateral triangles with parallel sides and common centroid (Figure 8.6), then the distance $d(M_1, M_2)$ is equal to $h\sqrt{3}$, where h is the distance between the corresponding parallel sides of the triangles. Indeed, the h-extension of the smaller triangle does not contain the vertices of the larger one.

Let now M_1, M_2, ... be a sequence of compact figures in the plane and M be one more compact figure. The figure M is said to be the *limit* of the sequence M_1, M_2, ... if the distance $d(M_1, M_k)$ approaches 0 *as k* increases without bound. In this case we also say that the sequence M_1, M_2, ... *converges* to M.

A sequence M_1, M_2, ... of compact figures in the plane is called *bounded*, if there exists a square that contains *all* the figures M_1, M_2, ... **The following theorem has important applications in mathematics.** It is, in a sense, the *Bolzano-Weierstrass Theorem for Compact Figures*.

Theorem 8.1. *Each bounded sequence of compact figures in the plane (or in the space R^n) has a convergent subsequence.*

Proof. Have you forgotten the pigeonholes? Let M_1, M_2, ... be a bounded sequence of compact figures, and m be a positive integer, such that all the figures M_1, M_2, ... are contained in the square with vertices $(\pm m, \pm m)$. We divide this square into unit squares (Figure 8.1) and presume that each unit square contains its boundary. Now each union of several unit squares we call a pigeonhole. So, there is a finite number of pigeonholes (more precisely, there are $4m^2$ unit squares and, consequently, 2^{4m^2} pigeonholes, including the empty pigeonhole, that is the "union" of the empty set of squares). Certainly, our compact sets M_1, M_2, ... are the pigeons! But in what sense is a pigeon sitting on a pigeonhole?

Let F be a compact figure contained in the square with the vertices $(\pm m, \pm m)$. We denote by $C_1(F)$ the union of the unit squares that have at least one common point with F. The figure $C_1(F)$ is said to be the *container* of F.

Now we say that the pigeon M_k is sitting on a pigeonhole if this pigeonhole is the container of M_k. So, each pigeon is sitting

80

on *only one* pigeonhole. In Figure 8.7 the pigeon M_k and the pigeonhole on which it is sitting are drawn.

Thus, we have a finite number of pigeonholes and infinitely many pigeons. Therefore, by the infinite pigeonhole principle, there exists a pigeonhole P_1 that contains infinitely many pigeons. We pick one of these pigeons and denote it by N_1. Obviously, $C_1(N_1) = P_1$. It is not difficult to show that if F is a compact figure with $C_1(F) = P_1$, then

$$d(N_1, F) \leq \sqrt{2}$$

(since the length of the diagonal of the unit square is equal to $\sqrt{2}$).

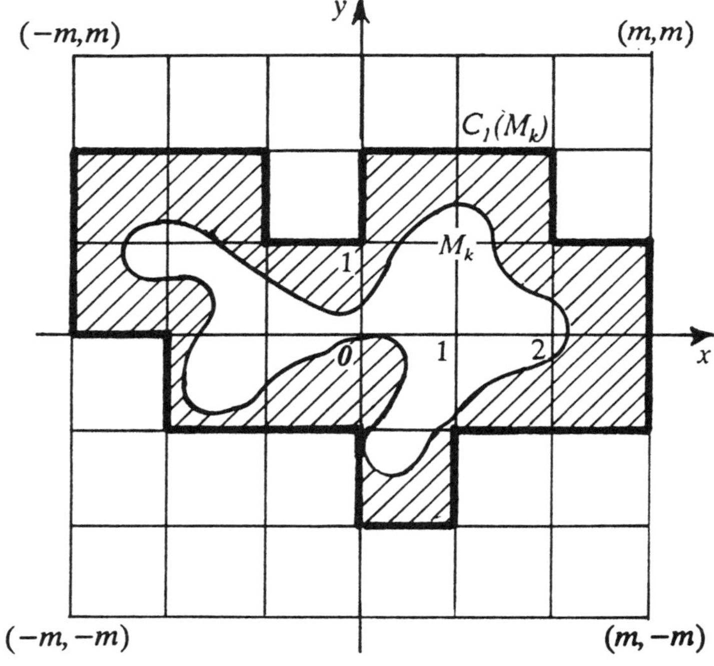

Figure 8.7

We now divide each unit square that is contained in P_1 into four squares of side $\frac{1}{2}$ (Figure 8.8). If M_k is a pigeon whose container coincides with P_1, then by $C_2(M_k)$ we denote the union

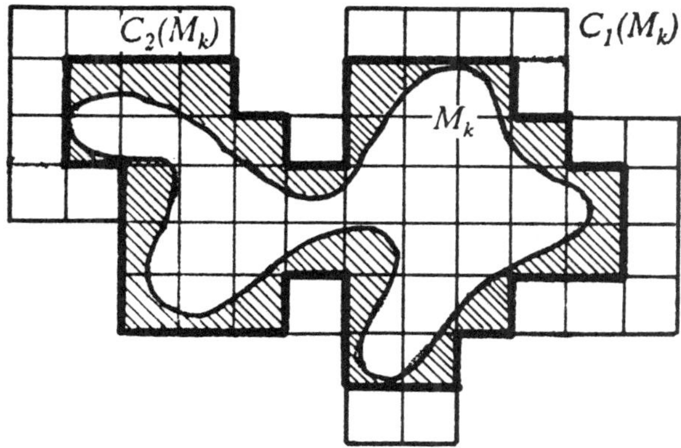

Figure 8.8

of all small squares (with side length $\frac{1}{2}$) that have at least one common point with M_k. The figure $C_2(M_k)$ is the new pigeonhole on which the pigeon M_k is sitting. Again, we have a finite number of pigeonholes where as the remaining flock of pigeons (for which $C_1(M_k) = P_1$) is infinite. Consequently, there exists a pigeonhole P_2 that contains infinitely many pigeons.

Out of these infinitely many pigeons sitting on P_2 we pick one, call it N_2, such that the index of N_2 in the original sequence $M_1, M_2, \ldots$ is greater than the index of N_1. Of course, $C_2(N_2) = P_2$. As in the first step, if F is a compact figure with $C_2(F) = P_2$, then

$$d(N_2, F) \leq \frac{\sqrt{2}}{2}$$

Then we divide each square of P_2 into four squares with side length $\frac{1}{4}$, and so on.

We obtain an infinite sequence of more and more "narrow" pigeonholes $P_1, P_2, \ldots$, and an infinite sequence $N_1, N_2, \ldots$ of distinct pigeons that is a subsequence of the given sequence $M_1, M_2, \ldots$ of compact figures. According to the construction, for each pigeon N_k of the created subflock we have $C_1(N_k) = P_1$; for each N_k with k $\geq$ 2 we have $C_2(N_k) = P_2$, and so on. This means that

$$d(N_1, N_k) \le \sqrt{2} \text{ for all } k > 1,$$

$$d(N_2, N_k) \le \frac{\sqrt{2}}{2} \text{ for all } k > 2,$$

and generally,

$$d(N_j, N_k) \le \frac{\sqrt{2}}{2^{j-1}} \text{ for all } k > j.$$

Finally, we denote by N the intersection of all pigeonholes P_1, P_2, ... Then N is a non-empty compact figure (see Example 8.4). It can be easily shown that $C_1(N) = P_1$, $C_2(N) = P_2$, ... Consequently, $d(N_j, N) \le \frac{\sqrt{2}}{2^{j-1}}$. This means that the sequence N_1, N_2, ... converges to N. The proof is completed.

Example 8.5. Mathematicians of ancient Greece knew that *among all figures of given perimeter P the circle has the maximal area.* They knew it but could not prove it! A nice proof was suggested in the Nineteenth Century by the Swiss mathematician Jacob Steiner. But his proof had a hole. We are going to discuss here the wonderful idea of Steiner and fill in the hole in his proof. This will serve as a good example of an application of the above Theorem 8.1. The problem of determining the figure F of maximal area with given perimeter P is called the *isoperimetric problem.* In this case, the figure F is called *extremal.*

Let F be a figure with the given perimeter P. If it is not convex, then we slide a rubber loop on it (Figure 8.9). We obtain

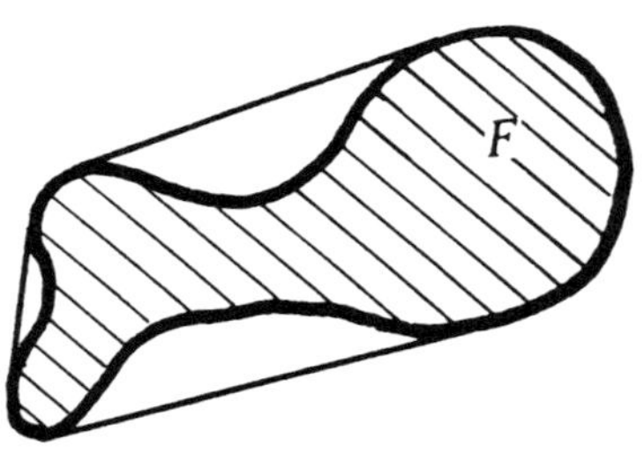

Figure 8.9

a figure of a larger area and smaller perimeter (we'll discuss convex figures in a greater detail in Chapter IV). So, if a figure is not convex, then it is not extremal, that is, it cannot be a solution of the isoperimetric problem.

Let now *F* be a convex figure. A chord $[a,b]$ of *F* we will call a ***cross-cut*** if it divides the boundary of *F* into two arcs of equal length $\frac{1}{2}P$. Steiner noticed that if *F* were extremal, then every cross-cut would divide its area into two *equal* parts. Indeed, if $S_1 > S_2$ (Figure 8.10), then by replacing S_2 with the symmetric image of S_1 we would obtain a figure with the same perimeter *P* and a *greater* area, contradicting the extremality of *F*.

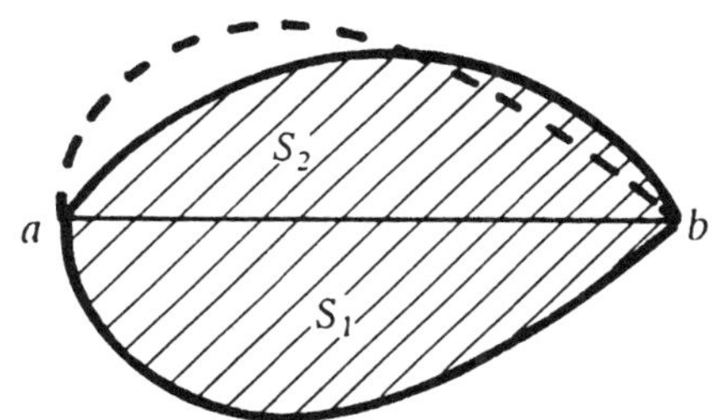

Figure 8.10

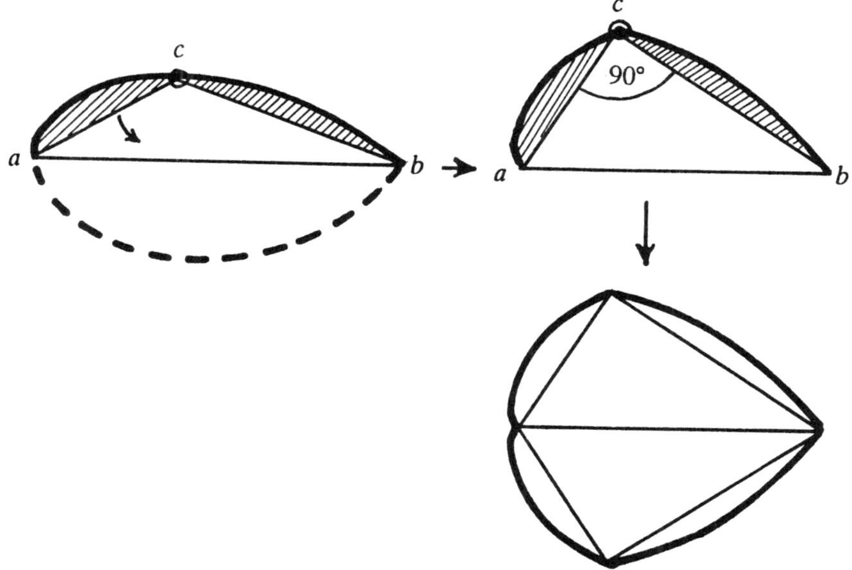

Figure 8.11

Then Steiner fixed a cross-cut $[a,b]$ of the extremal figure F and showed that for any boundary point c of F distinct from a and b, the angle acb must be equal to $90°$. Indeed, if the angle abc were not equal to $90°$ (Figure 8.11), then we could install a hinge in the point C and rotate one of the shaded areas about c until the angle acb is equal to $90°$ (Figure 8.11). While doing so, we certainly would not change the shaded areas, but would increase the area of the triangle acb (prove it!). Now we can replace the lower half of F by a symmetric image of the upper half. Thus, we get a figure of the same perimeter as F but of a greater area.

But if for every boundary point c of F (except a and b) the angle acb is equal to $90°$, then F would have to be a disk (please see the definition of a disk on page 73).

May we now conclude that the disk of perimeter P has the maximal area among all figures of perimeter P? In order to answer this question, let us consider the following hypothetical situation.

A customer came to a clock shop and asked to have his antique clock with a very complicated mechanism repaired. The owner of the shop asked his experts, and they replied that nobody except the foreman Smith could repair the clock. Smith was ill so the customer was asked to come next week. However, the next week it became clear that the foreman Smith also was unable to repair that clock. So "nobody except Smith can" does not mean yet that "Smith can."

Similarly, the Steiner's reasoning above shows that no figure except the disk can be extremal. Does it mean that the disk is surely the extremal figure? Not at all! Perhaps no figure except the disk is extremal, but the disk is not extremal too. If we had a guarantee for the *existence* of an extremal figure, then (knowing that no figure except the disk is extremal) we could conclude that the disk *really is* the extremal figure. The lack of such a guarantee of existence is the hole in Steiner's reasoning.

And what is a guarantee for existence of an extremal figure? Theorem 8.1 gives such a guarantee! Indeed, let us denote by S the exact upper bound of areas of all figures with perimeter P. Then for every positive integer k there exists a figure M_k of perimeter P whose area is greater than $S - \frac{1}{k}$. We can place all the figures M_1, M_2, ... in a bounded piece of the plane (say, in a circle of radius P). We obtain the bounded sequence M_1, M_2, ... of compact figures in the plane. By virtue of Theorem 8.1 there exists a convergent subsequence. Now it is not difficult to prove that the limit figure N of this subsequence has area S, that is, the figure N is extremal. This is a way to fill in the hole in Steiner's reasoning. So, the disk *is* the only extremal figure.

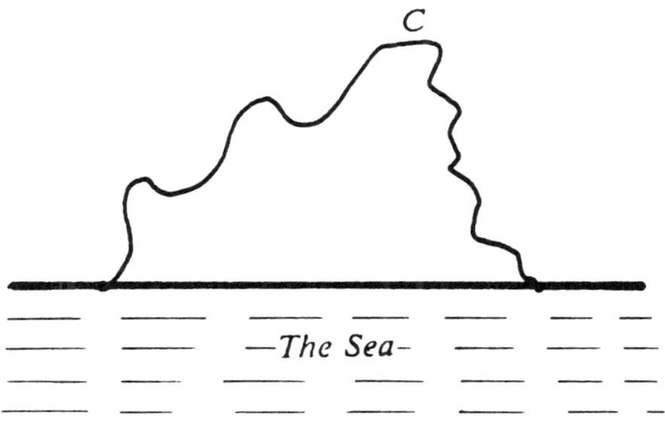

Figure 8.12

Exercise 8.4. According to a legend, the Queen Dido permitted a town to be built by the sea "bounded by an ox's skin." The skin was cut into thin strips and then the strips were tied into a long ribbon. But how were they to bind the region of maximal area with the help of this ribbon (Figure 8.12)? Can you help the people in this legend?

Exercise 8.5. (B. Grünbaum; the same as Exercise 6.9) Prove that every tiling of an infinite strip by copies of a polyominal tile T contains a bounded part that can be used to tile a cylinder.

Solutions of Exercises

8.1. Let us denote by x the distance between the given line L and the constructed line L_1 cutting F into two parts (Figure 8.4). We denote by S_1 and S_2 the areas of these parts. Then the difference $S_1 - S_2$ is a continuous function of x (we don't give a proof of this visually clear assertion, because we did not introduce the exact definition of continuity). But when x is small the difference $S_1 - S_2$ is equal to S (Figure 8.4), and when x is large $S_1 - S_2 = -S$. So, according to the Intermediate Value Theorem, there is a value $x = c$ for which $S_1 - S_2 = 0$, that is, $S_1 = S_2 = \frac{1}{2}S$.

8.2. Let L be a directed line through p that forms an angle ϕ with a fixed initial ray L_0 (Figure 8.13). We denote by S_1 and S_2 the areas of the parts into which L divides (S_1 is the area of the part to the left from L). So, the difference $S_1 - S_2$ is a continuous

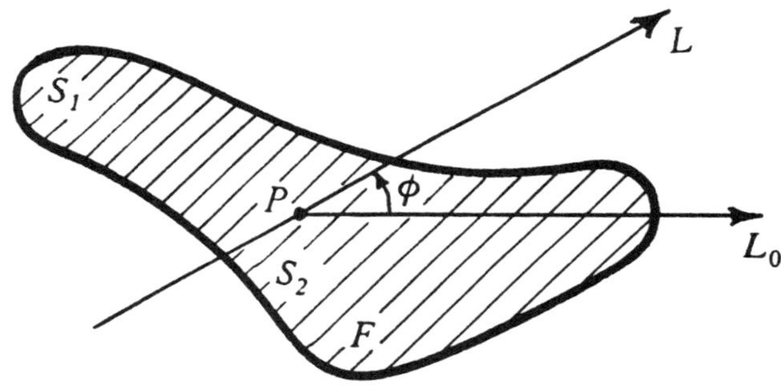

Figure 8.13

function of ϕ. But when ϕ runs all the values from 0 to π, the areas S_1 and S_2 interchange their roles. This means that if $S_1 - S_2$ was negative for $\phi = 0$, then it would be positive for $\phi = \pi$. Consequently, by virtue of the Intermediate Value Theorem, there exists a value $\phi = c$ for which $S_1 - S_2 = 0$, that is $S_1 = S_2 = \frac{1}{2}S$.

8.3. We squeeze the figure F between two parallel lines L_1 and L_2, that form an angle ϕ with a fixed initial ray (Figure 8.14). Then we squeeze F between two parallel lines M_1 and M_2 that form the angle $\phi + 90°$ with the initial ray (Figure 8.15). The four lines define a rectangle circumscribed about F. Let a be the length of its side parallel to L_1 and b be the length of the side parallel to M_1. Then the difference $a - b$ is a continuous function of ϕ. But when ϕ runs through the values from 0 to π, the lengths a and b interchange their roles. This means that if $a - b$

> 0 for $\phi = 0$, then $a - b < 0$ for $\phi = \pi$. Consequently, there exists an angle $\phi = c$ for which $a - b = 0$, that is, the circumscribed rectangle turns into a square.

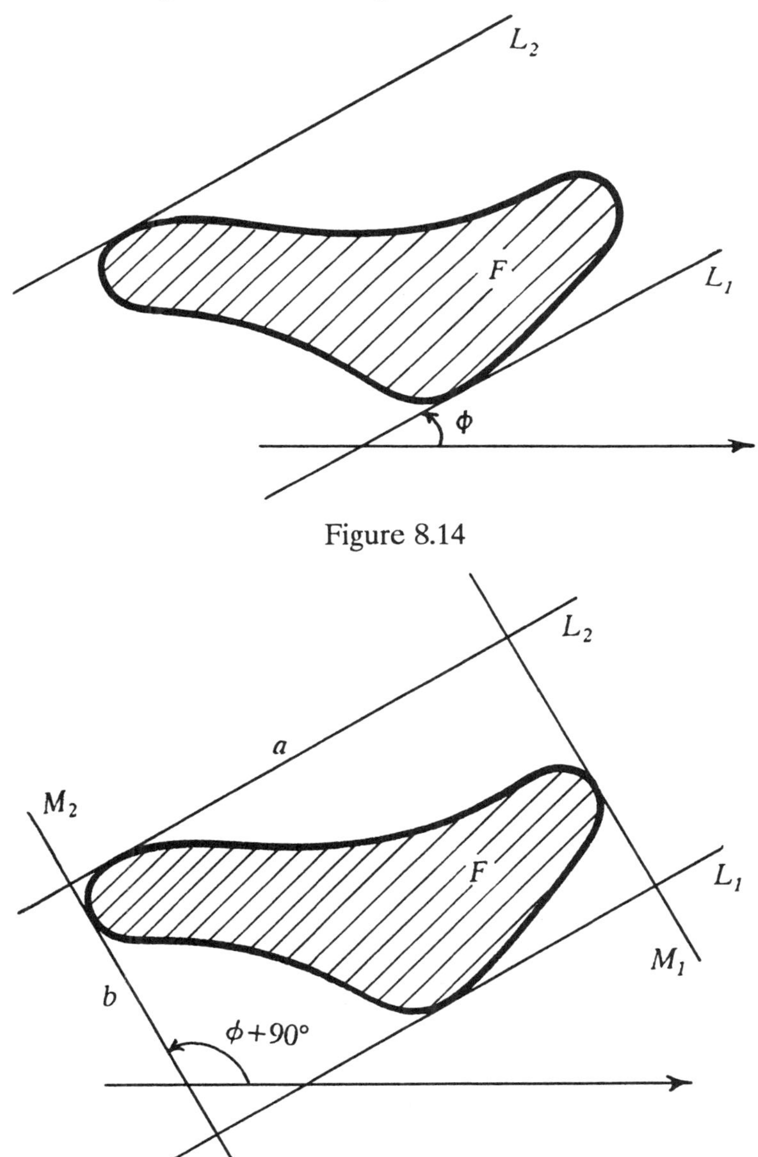

Figure 8.14

Figure 8.15

8.4. Let C be a curve bounding the town (Figure 8.12), q its length, and C' the curve, that is symmetric to C with respect to the sea line L (see Figure 8.16). Then the union of C and C' is a closed curve of the length $2q$. We obtain the maximal area when this union is a circle; hence, C is a semicircle (Figure 8.17).

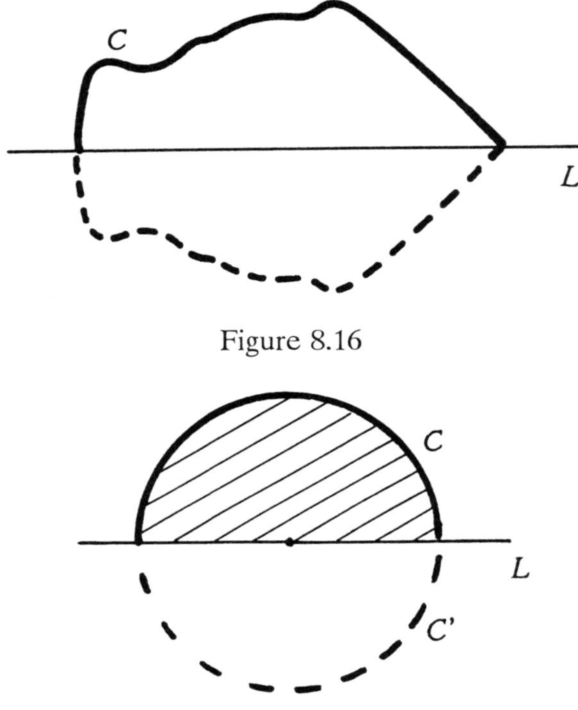

Figure 8.16

Figure 8.17

8.5. Let S be an infinite strip tiled by copies of a tile T. There are only finitely many shapes of step-lines (can you prove it?) cutting across S along the boundaries of tiles (Figure 8.18 shows one such step-line cut).

On the other hand, there are infinitely many step-line cuts because S is an infinite strip. By the Infinite Pigeonhole Principle there is a shape of a cut that repeats at least twice. We get a region F just like the one in Figure 6.5. You can glue a cylinder out of F.

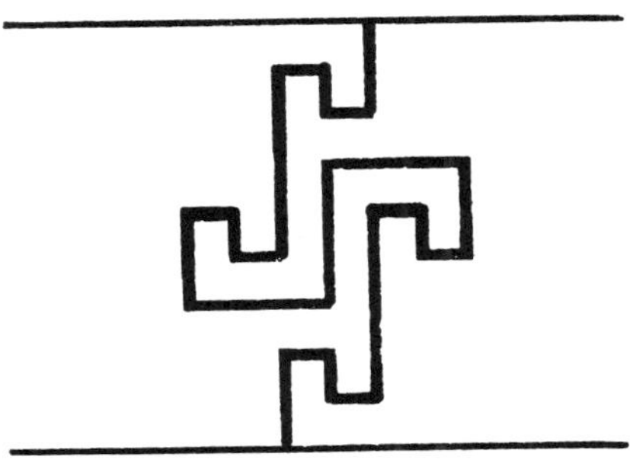

Figure 8.18

CHAPTER 3

A WORD ABOUT GRAPHS

9. COMBINATORICS OF ACQUAINTANCE, OR INTRODUCTION TO GRAPH THEORY

In mathematics it is sometimes possible to derive something out of virtually nothing.

Example 9.1. A number of people (more than one) come to a party. Prove that at least two of them have an equal number of acquaintances in the party. (The notion of acquaintance is reflexive: if A is acquainted with B, then B is acquainted with A.)

Solution. If n people came to the party, then for each of them the number of acquaintances is an integer ranging from 0 to $n - 1$. In fact, 0 and $n - 1$ may not both serve as numbers of acquaintances for people in our party because 0 implies existence of a person not acquainted with anybody whereas $n - 1$ claims a person acquainted with everyone.

Thus, we have $n - 1$ numbers that may serve as numbers of acquaintances, and n people in the party. The Pigeonhole Pinciple now proves the required result.

In the solution of Example 9.1 we discovered that at most $n - 1$ integers can appear as numbers of acquaintances: 0, 1, ..., $n - 2$ or 1, 2, ..., $n - 1$. Our natural curiosity prompts the following question.

Exercise 9.1. Is there a party of n such that

a)   every integer 0, 1, ..., $n - 2$ appears as the number of acquaintances of a person in the party?

b) every integer 1, 2, ..., $n - 1$ appears as the number of acquaintances of a person in the party?

Exercise 9.2. Can it so happen that every person has exactly 7 acquaintances in a party of 21?

If we know that at least six people came to a party, we can prove another exciting result that is due to the English mathematician Frank Ramsey.

Example 9.2. Prove that in any party of six people there are three mutual acquaintances or three mutual non-acquaintances.

Solution.

(a) It is convenient to record the information about acquaintances of a group of six people by a combination of two diagrams G and $\overline{G}$. In both diagrams we represent every person by a vertex for the total of six vertices, and two vertices of G are connected by a curve (curve shape is irrelevant) if and only if they correspond to two people who are acquainted.

Two vertices of $\overline{G}$ are connected by a curve if and only if they correspond to two people who are not acquainted.

Please note that any two vertices are connected in exactly one of the two diagrams G, $\overline{G}$.

The given problem is now equivalent to proving that at least one of the two diagrams G, $\overline{G}$ contains a triangle (i.e., three pairwise connected vertices)!

(b) Let us fix the same vertex A in both diagrams G and $\overline{G}$. A is connected with each of the five other vertices either in G or in $\overline{G}$, therefore, it must be connected with at least three of the five vertices in G or in $\overline{G}$ (Pigeonhole Principle with five pigeons and two holes).

Due to the symmetry of the problem, we can assume without loss of generality that A is connected with the vertices B_1 B_2, and B_3 in G. If any two of the vertices B_1, B_2, B_3 are connected in G, then these two vertices and A are the vertices of a triangle contained in G (Figure 9.1).

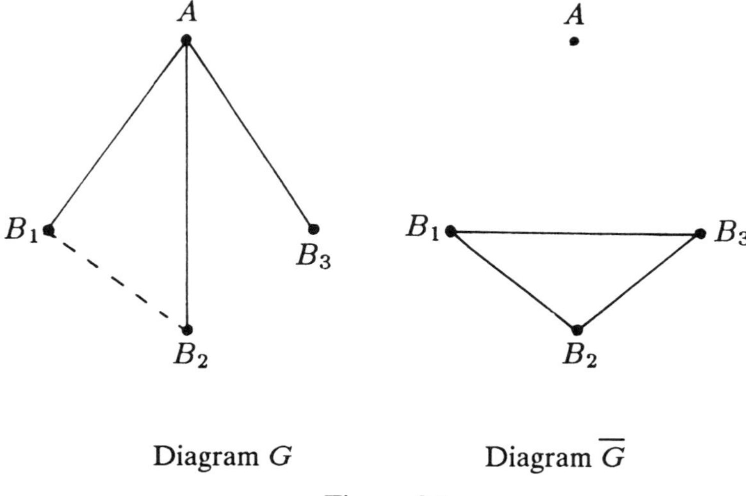

Diagram G Diagram $\overline{G}$

Figure 9.1

If no two of the vertices B_1, B_2, B_3 are connected in G, then the triangle B_1, B_2, B_3 is contained in the diagram $\overline{G}$.

There is another way to look at the party-of-six problem. Given six vertices, no three on a line, every two of them are connected by an edge. We get the complete network with 6 vertices (Figure 9.2).

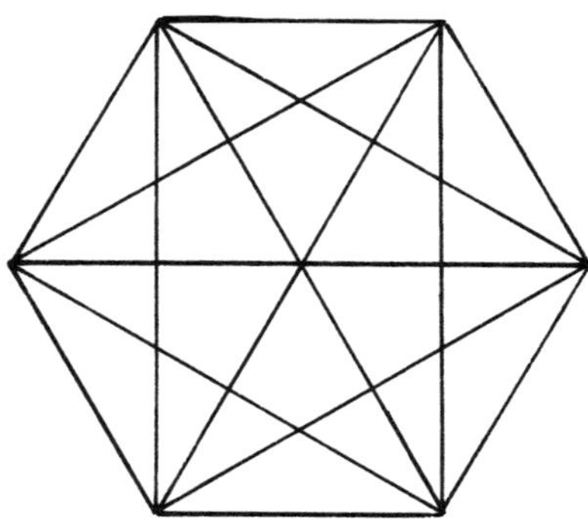

Figure 9.2

The party-of-six statement is equivalent to the following: *no matter how the edges of the complete network with 6 points are colored in two colors (each edge in one color), there is always a one-color triangle (with vertices in the given points).*

Exercise 9.3. Prove that six is the smallest number of people in the party to guarantee the result of Example 9.2, i.e., party of five people might have neither three mutual acquaintances nor three mutual non-acquaintances.

Exercise 9.4. Prove that in any party of nine there are three mutual acquaintances or four mutual non-acquaintances.

Exercise 9.5. Prove that nine is the smallest number of people in the party to guarantee the result of Exercise 9.4.

A *cycle of acquaintances* is a party of n people, such that the first and the second persons are acquainted, the second and the third persons are acquainted, etc., the nth and the first persons are acquainted. (Other acquaintances among these n people may be present, but they do not interest us in this definition.)

Exercise 9.6. Prove that any party of people that contains at least two people, in which any cycle of acquaintances consists of an even number of people, can be partitioned into two non-empty groups, such that either group consists of mutual non-acquaintances.

Exercise 9.7. Prove the converse of the statement of Exercise 9.6.

In case you have not noticed, on the pages above we have introduced you to an area of combinatorial mathematics called Graph Theory.

A *graph* is a finite non-empty set V of **vertices** some pairs of which (or perhaps none) are said to be adjacent. An adjacent pair $e = \{v_1, v_2\}$ of vertices is called an **edge**. In this case we say that e and v_1 are **incident**, as are e and v_2.

The diagram G of acquaintances for a party is a good example of a graph.

If it so happened that every vertex of a graph G_1 is also a vertex of a graph G, *and* every edge of G_1 is also an edge of G, then the graph G_1 is said to be a **subgraph** of the graph G.

The diagram $\overline{G}$ of non-acquaintances is uniquely determined by the graph of acquaintances G, and is called the **complementary graph** of the graph G or simply the **complement** of G.

The number of edges incident to a vertex v in a graph is called the **degree** of v and is denoted as *deg v*. The following statement can be proved by reasoning identical to the one we used to solve Exercise 9.1. The symbol $\sum_{v \in V} deg\ v$ in it denotes the sum of all numbers *deg v* where $v \in V$ (i.e., v runs through all elements of the set V).

Theorem 9.1. *Prove that for any graph G*

$$\sum_{v \in V} deg\ v = 2q$$

where V is the set of vertices of G, and q is the number of edges.

If the vertex set V of a graph G can be partitioned into two non-empty subsets V_1 and V_2 such that no edges connect vertices in either of the subsets (but may connect points from V_1 to points from V_2), then F is called a **bigraph** or *bipartite graph*.

Exercises 9.5 and 9.6 provide a nice description of bigraphs. In the language of graph theory it sounds as follows:

Theorem 9.2. *A graph G is a bigraph if and only if it contains no odd cycles.*

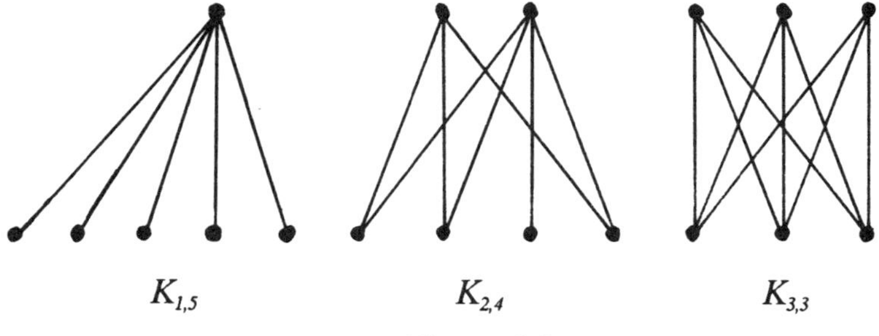

Figure 9.2

Let G be a bigraph with the vertex set V partitioned into subsets V_1 and V_2 with n_1 and n_2 elements respectively. If *every* vertex of V_1 is adjacent to *every* vertex of V_2, then G is called a **complete bigraph** and denoted by K_{n_1,n_2}. Obviously, K_{n_1,n_2} contains $n_1 n_2$ edges. Figure 9.2 gives examples of all complete bigraphs with six vertices.

Exercise 9.8. Let $n = n_1 + n_2$ and $n = m_1 + m_2$ be two distinct decompositions of n into the sum of two positive integers (i.e., $n_1 \neq m_1$ and $n_1 \neq m_2$). Is there a graph G such that $G = K_{n_1,n_2}$ and at the same time $G = K_{m_1,m_2}$?

A graph corresponding to a party of n mutual acquaintances is called a **complete graph** and is denoted by K_n.

For positive integers m and n, the **Ramsey number** $r(m,n)$ is the smallest positive integer R such that for any graph G with R vertices, G contains K_m as a subgraph or $\overline{G}$ contains K_n as a subgraph.

Now we are ready to translate Example 9.2 and Exercises 9.2, 9.3, and 9.4 into the language of graph theory.

Theorem 9.3. $r(3,3) = 6$; $r(3,4) = 9$

Of course, the function $r(m,n)$ is symmetric:

Exercise 9.9. For any positive integers m and n

$$r(m,n) = r(n,m)$$

Now, over sixty years after the pioneering paper by Frank Ramsey, we know only a few small Ramsey numbers. They all are presented in the following table:

n \ m	2	3	4	5	6	7	8	9	...
2	2	3	4	5	6	7	8	9	...
3	3	6	9	14	18	23		36	
4	4	9	18						

The dots in the table suggest the equality $r(2,n) = n$.

Exercise 9.10. Prove that for any positive integer n,

$$r(2,n) = n$$

But even $r(4,5)$ or $r(5,5)$ are not known. So if you find new partial solutions of the following problem be sure to write to us!

Problem 9.1. *For positive integers m and n find $r(m,n)$.*

It is a very difficult problem. One of the great mathematicians of this century, Paul Erdös, told the following story about it. Aliens invade the earth and threaten to destroy it in a year if human beings would not find *r(5,5)*. It is, probably, possible to save the earth by putting together the world's best minds and computers. If, however, the invaders were to demand *r(6,6)*, the human beings might as well attack first without even trying to ponder the problem.

Solutions of Exercises

9.1. (a) We can prove by induction on n that there is a party of n such that every integer 0, 1, ..., $n - 2$ appears as the number of acquaintances of a party member.

Two non-acquainted people solve it for $n = 2$.

Assume that there is a party P of n such that every integer 0, 1, ..., $n - 2$ appears as the number of acquaintances. In fact, by the pigeonhole principle one of these numbers say k ($0 \leq k \leq n - 2$) must appear twice.

Now we have to construct a party P' of $n + 1$, such that every integer 0, 1, ..., $n - 1$ appears as the number of acquaintances. We start with the party P of n and add one more person, who is acquainted with exactly one of two people with k acquaintances and with everyone having more than k acquaintances. You can easily verify that the party P' satisfies the required condition.

(b) There is such a party. The construction is similar to the one in Exercise 9.1(a).

9.2. One relation of acquaintance, say between people A and B, adds one unit to the number of acquaintances of each A and B, i.e., it adds two units to the total number T of acquaintances of the party. Therefore T is even.

On the other hand, if we assume that a party of 21 exists such that everyone has exactly seven acquaintances, we will have $T = 21 \times 7$, an odd number. This contradiction proves that such a party does not exist.

We would like to mention here that similar reasoning proves the following result: *there is no polyhedron with an odd number of odd-sided faces.*

9.3. See:

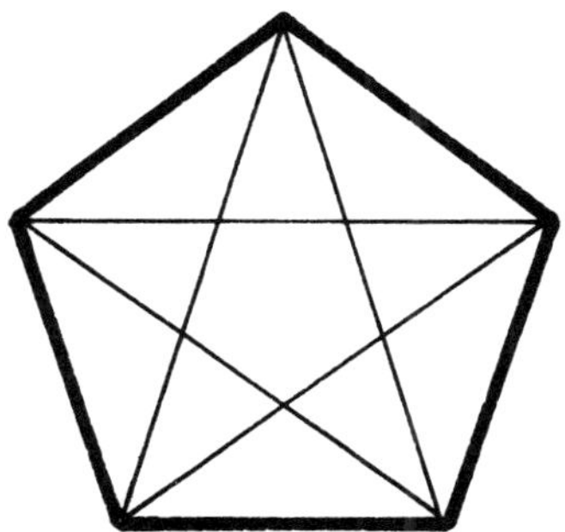

Figure 9.3

9.4. Assume there is a person, say A, acquainted with at least four party members, say B_1, B_2, B_3, and B_4 (Figure 9.4).

Then should at least two of the B's, say B_1 and B_2, be acquainted, we get a triangle AB_1B_2 of mutual acquaintances. Otherwise B_1, B_2, B_3, and B_4 are mutually non-acquainted.

Now consider the opposite case, i.e., every person is acquainted with not more than three party members. This means exactly that everyone has at least five non-acquaintances. Now, there can not be a party of nine such that every party member has exactly five non-acquaintances (because the total number of non-acquaintances must be even, where as 9×5 is odd -- reasoning

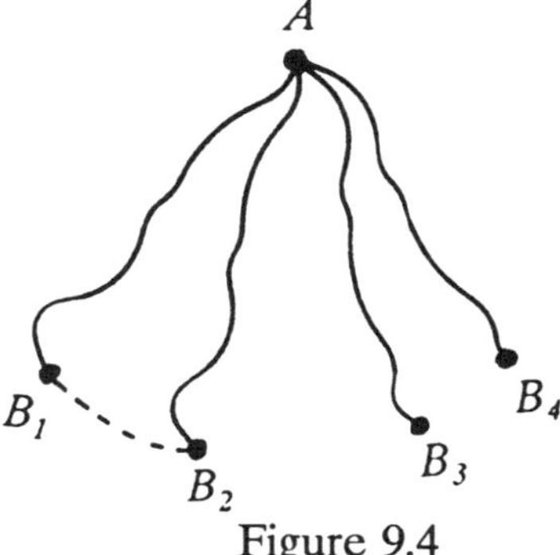

Figure 9.4

identical to the solution of Exercise 9.2). Therefore, at least one party member, say A, has at least six non-acquaintances B_1, B_2, ..., B_6 (Figure 9.5).

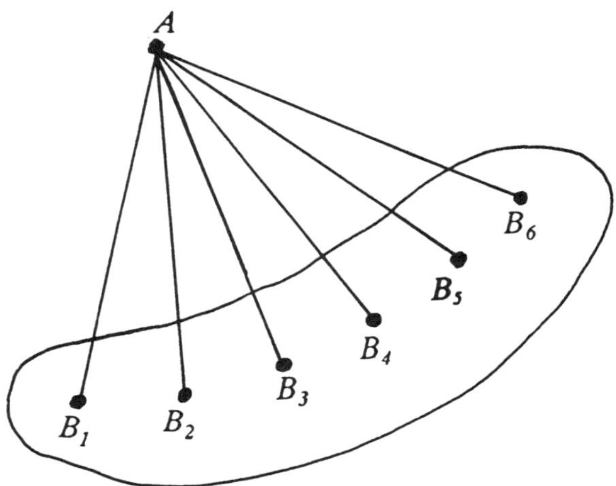

Figure 9.5

All there is left to do is to apply the party-of-six result (Example 9.2) to the party of B_1, B_2, ... B_6. If three of them are mutual acquaintances, then we are done. If three of them are mutual non-acquaintances, then three of them and A form four mutual non-acquaintances.

No doubt you noticed that the party-of-nine result is not symmetric; three acquaintances -- four non-acquaintances, and there is nothing we can do about it. The good news is, however, that we can formulate a dual result that will be true too (do you see why?). *In any party of nine there are four mutual acquaintances or three mutual non-acquaintances.*

9.5. See:

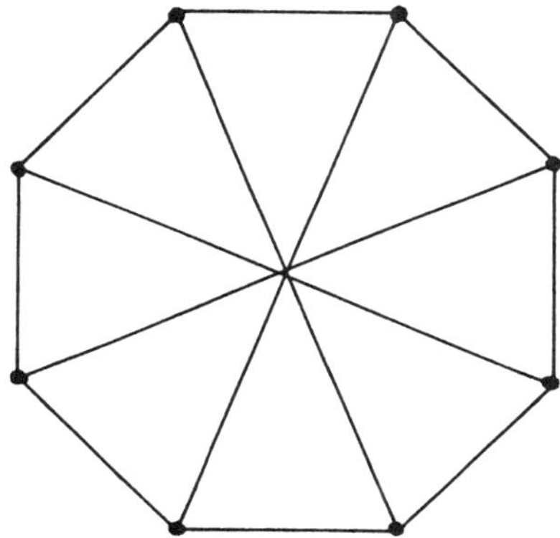

Figure 9.6

9.6. As in the previous problems, it is convenient to represent people by vertices and connect "acquainted" points by curves. We need to partition the party into two subsets S_1 and S_2. We pick a vertex v and assign it to S_1. Now every vertex from which you can walk to v through a number of curves can be assigned the distance:

$$d(s,v) = \begin{cases} 0 \text{ if you walk through an even number of edges} \\ 1 \text{ if you walk through an odd number of edges} \end{cases}$$

This distance is uniquely defined because if we assume that one even and another odd walks exist from any w to v (Figure 9.7), then it is easy to show (please do) the existence of an odd cycle of acquaintances.

Now we assign the points w of distance 0 from v to S_1, and the points of distance 1 to S_2.

If any point v_1 is not assigned yet, we assign v_1 and all points of distance 0 from v_1 to S_1, and the points of distance 1 from v_1 to S_2, etc (we only have finitely many points).

Figure 9.7

Now we have to prove that two points x,y from the same subset, say S_1, cannot be adjacent. Indeed, if we assume the opposite, we can show (please do) the existence of an odd cycle of acquaintances (Figure 9.8).

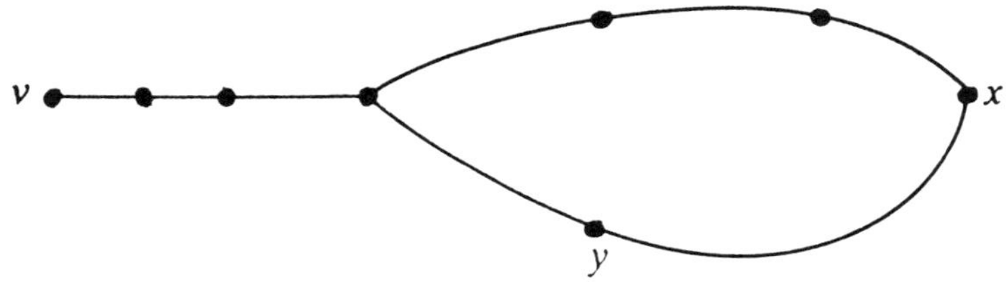

Figure 9.8

9.7. Once again we represent people by vertices and adjacency between them by curves, and assume that the set of vertices is partitioned into two non-empty subsets S_1 and S_2 of mutual non-acquaintances (Figure 9.9).

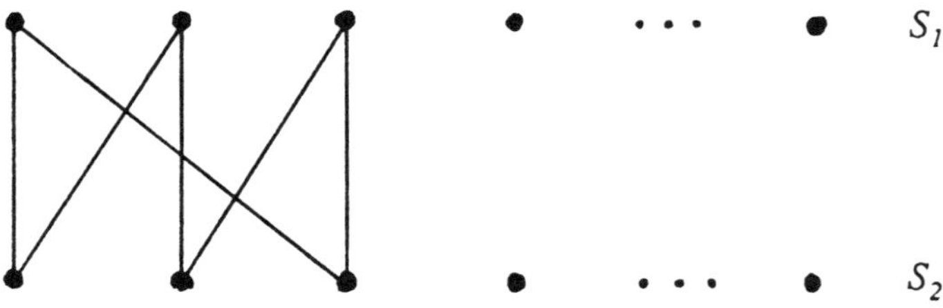

Figure 9.9

The points in a cycle must alternate: one from S_1, next from S_2, then from S_1 again, etc. This shows that any cycle of acquaintances must be even.

9.8. Assume that such a graph G exists. Then two distinct decompositions of n (n is the number of vertices of G) guarantee that when we present G as K_{n_1, n_2} and K_{m_1, m_2}, the corresponding decompositions $V = V_1 \cup V_2$ and $V = W_1 \cup W_2$ of the vertex set V of G are distinct. In other words, there are two vertices x and y in V, such that both x and y belong to, say, V_1, and at the same time x is an element of W_1 where as y is an element of W_2. But this means that x and y are not adjacent, and at the same time x and y are adjacent. This contradiction proves that such a graph G does not exist.

9.9. Every graph G can be viewed as the complement of its complement $\overline{G}$.

9.10. Let G be a graph with n vertices. Then if any two points of G are adjacent, G contains a subgraph K_2. If no points of G are adjacent, then $\overline{G} = K_n$. Therefore $r(2,n) = n$.

10. MORE ABOUT GRAPHS

Graphs appear in our discussion as diagrams of acquaintances. Thus, the only things that matter when we represent graphs in the plane are the set of vertices (but not their positions on the plane as we draw them) and which vertices are adjacent (but not the shapes of edges, which we presume have no points in common, except vertices of the graph). In fact, think of a graph as a set of pins, some of which are connected by rubber bands. So a graph remains the same if we reposition the pins and stretch the rubber bands.

Two graphs are called *isomorphic* if "pins" of one of them can be repositioned and its "rubber bands" stretched so that this graph becomes graphically identical to the other graph.

More formally, two graphs, G and G_1 are said to be *isomorphic* if there is a one-to-one correspondence $f: V \rightarrow V_1$ of their vertex sets that preserves adjacency, i.e., vertices v_1 *and* v_2 of G are adjacent if and only if the vertices $f(v_1)$ and $f(v_2)$ of G_2 are adjacent.

We denote isomorphism of graphs G and G_1 by $G \cong G_1$.

Example 10.1. *Are the graphs in Figure 10.1 isomorphic?*

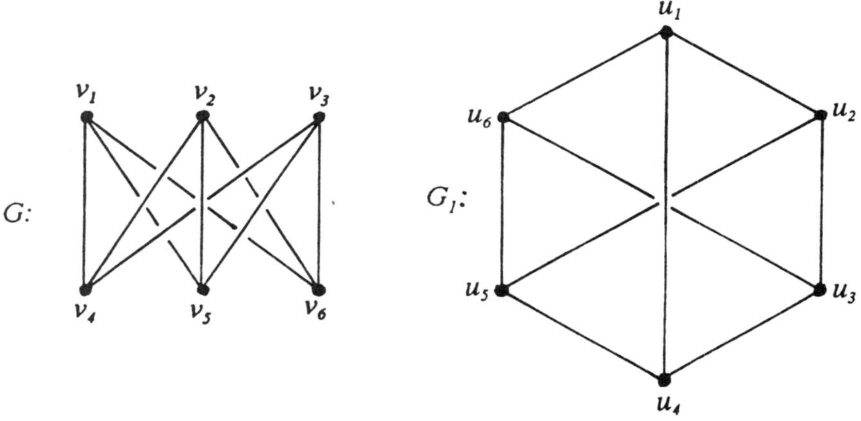

Figure 10.1

Solution. Let us manipulate "pins" and "rubber bands" of the graph G: first we flip $v_2 v_5$, then stretch it (Figure 10.2).

Lo and behold, we ended up with a graph graphically identical to the graph G_1. G and G_1 are isomorphic.

Needless to say, two isomorphic graphs *must* have equal numbers of vertices and edges; therefore, the numbers of vertices and edges are what we call **invariants** of graphs (invariants are characteristics shared by all isomorphic graphs). The equality of these two invariants, however, is not sufficient for isomorphism of two graphs.

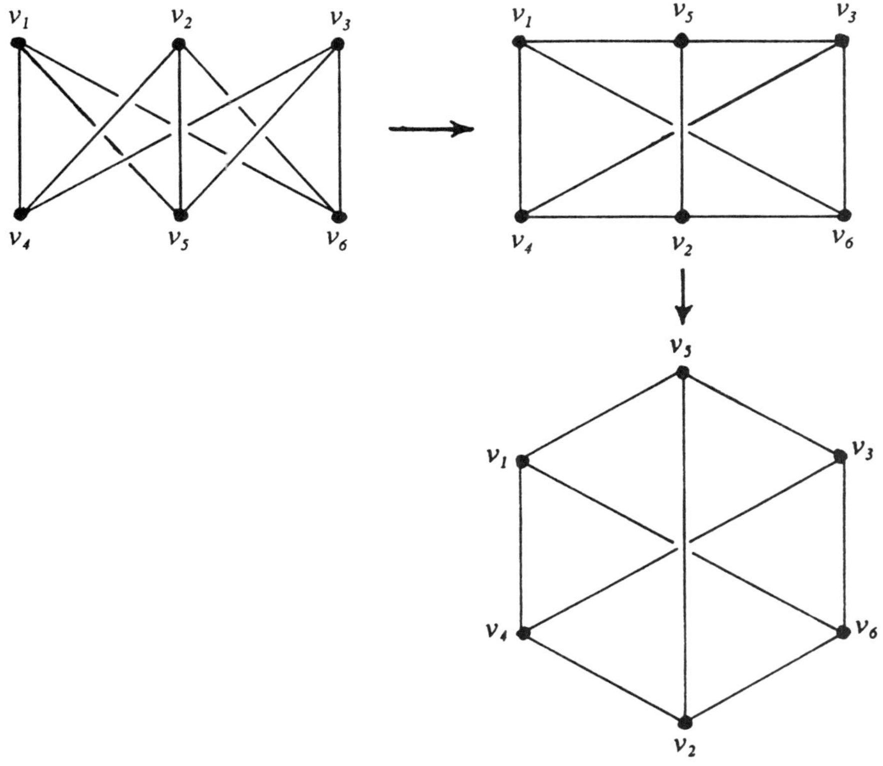

Figure 10.2

Example 10.2. *Are the graphs in Figure 10.3 isomorphic?*

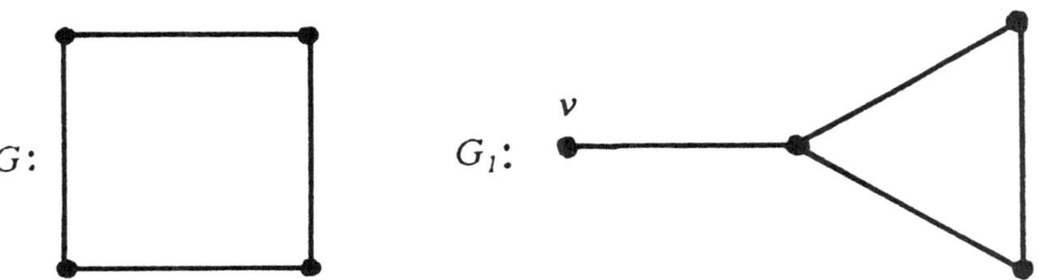

Figure 10.3

104

Solution. Even though the graphs G and G_1 have equal numbers of vertices and edges, they are not isomorphic. Under no one-to-one correspondence of vertices can adjacency be preserved because the graph G_1 contains a vertex v of degree one. This point must correspond under isomorphic correspondence to a point of degree one in graph G. But the graph G has no such point!

While solving Example 10.2, we discovered other invariants of graphs: degrees of the vertices. Do we have enough invariants to guarantee isomorphism of two graphs? In other words, given two graphs G and G_1 with equal numbers of vertices and edges, and equal sequences of vertex degrees, do G and G_1 have to be isomorphic?

Example 10.3. *Are the graphs in Figure 10.4 isomorphic?*

You may have sensed that any characteristic of a graph G that can be expressed in terms of adjacency, will be an invariant (i.e., will be shared by all graphs isomorphic to G). Such invariants include length of cycles, complete subgraphs, etc.

Now you can find many ways to prove that graphs G and G_1 are not isomorphic, even though their numbers of vertices, edges, and sequences of degrees are equal.

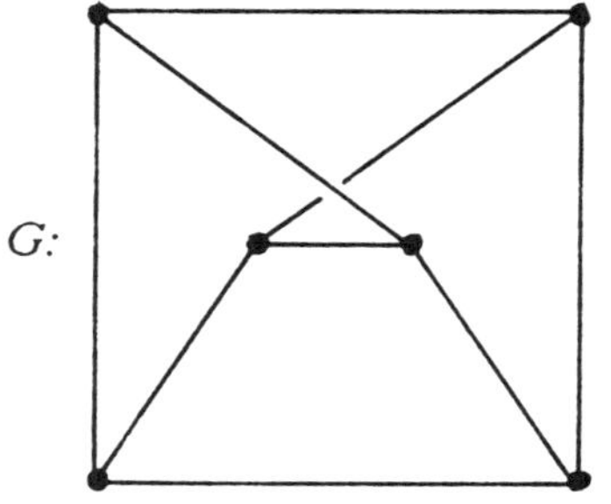
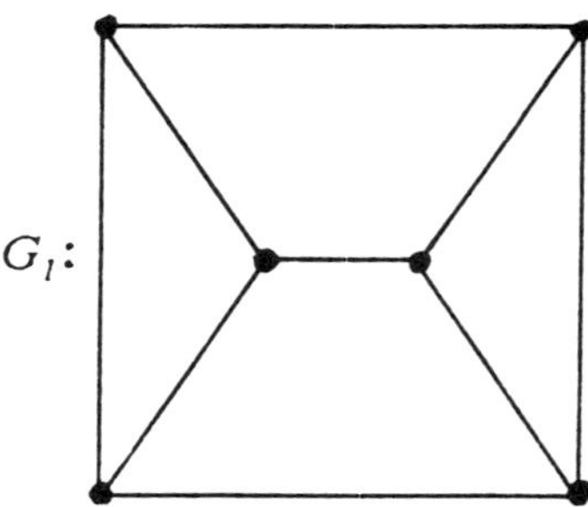

Figure 10.4

First Solution of Example 10.3. G and G_1 are not isomorphic because G_1 contains a triangle (three mutual acquaintances!) where as G does not.

Second Solution of Example 10.3. G and G_1 are not isomorphic because G is isomorphic to a bigraph $K_{3,3}$ where as G_1 is not a bigraph at all (prove both of these statements!).

As we were writing these lines for you, the following problem came to mind: *find all positive integers n such that there is a graph of order n that is isomorphic to its complement.*

Exercise 10.1. Is there a graph G that is isomorphic to its complement $\overline{G}$?

The *order* $|G|$ of a graph G is the number of vertices in G.

Exercise 10.2. Prove that if $G = \overline{G}$, then

$$|G| \equiv 0 \ (\mathrm{mod}\ 4)$$

$$\text{or}$$

$$|G| \equiv 1 \ (\mathrm{mod}\ 4)$$

Exercise 10.3. Is there a graph of order 4 that is isomorphic to its complement?

Exercise 10.4. Is there a graph of order 8 that is isomorphic to its complement?

Now we are ready to attack the following theorem.

Theorem 10.1. *There is a graph of order n that is isomorphic to its complement if and only if*

$$n \equiv 0 \ (\mathrm{mod}\ 4)$$

$$\text{or} \qquad\qquad (*)$$

$$n \equiv 1 \ (\mathrm{mod}\ 4)$$

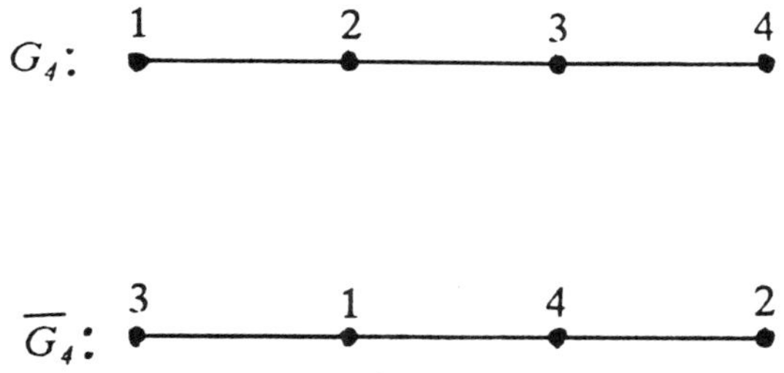

Figure 10.5

Proof. The necessity of condition (*) was proven in Exercise 10.2 above.

Let us prove the sufficiency.

For $n = 4$ the solution G_4 and its complement $\overline{G}_4$ are presented in Figure 10.5.

To construct G_8 we take two solutions for $n = 4$ and *sew* them together by a cycle of length 8 (Figure 10.6). The complement $\overline{G}_8$ (Figure 10.7) is graphically identical to G_8.

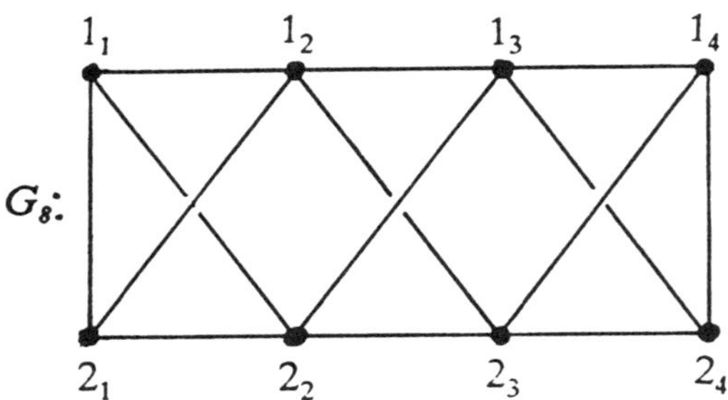

Figure 10.6

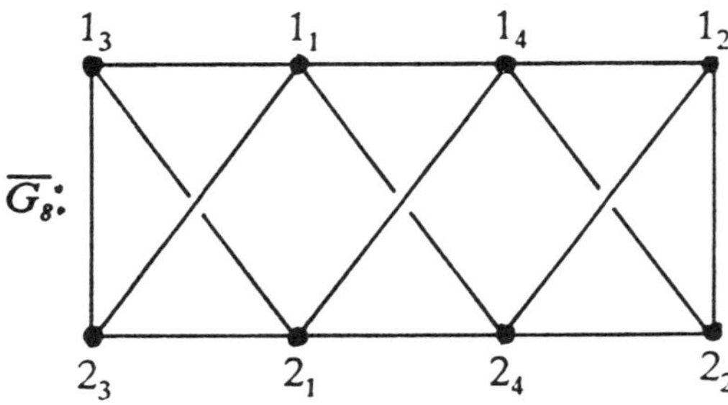

Figure 10.7

Now we can generalize this construction to create G_{4m} for any positive integer m.

In order to construct the complete graph K_m, we start with m vertices and connect every two of them together. Similarly we start with m copies of the graph G_4 (Figure 10.5) and *sew* every two of them together. We get the graph G_{4m} that is isomorphic to its complement $\overline{G}_{4m}$. We leave the proof of isomorphism to our readers.

Now let us look at the case when $n = 4m + 1$. For $m = 0$ the solution G_1 is trivial (Figure 10.8).

$$G_1: \quad \bullet$$

Figure 10.8

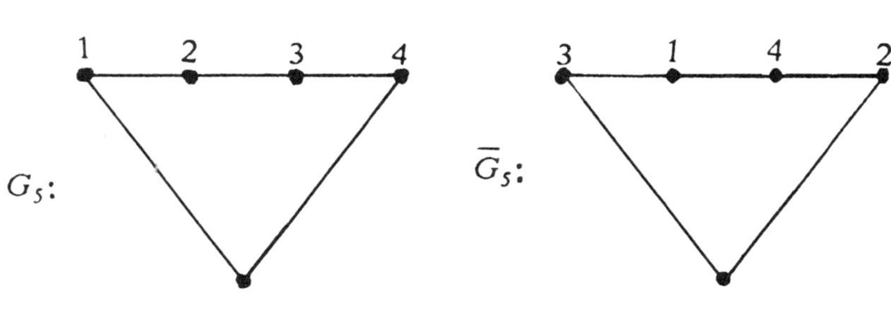

$G_5:$ $\overline{G}_5:$

Figure 10.9   Figure 10.10

When $m = 1$ we expand our *sew* operation to the situation when we wish to *sew* G_1 and G_4 together. The result is G_5 (Figure 10.9) that is isomorphic to its complement $\overline{G}_5$ (Figure 10.10).

Now you can guess how we are going to construct G_{4m+1} for any positive integer m. We start with m copies of the graph G_4 and one copy of the graph G_1 and *sew* every two of these $m + 1$ basic pieces together. We get G_{4m+1} that is isomorphic to its complement $\overline{G}_{4m+1}$. We would like our readers to prove this isomorphism on their own.

Exercise 10.5. Is there a graph G isomorphic to its complement $\overline{G}$, and not isomorphic to any of the graphs G_n constructed in our proof of Theorem 10.1?

Let G be a graph of order n. G and $\overline{G}$ appear when we color the edges of the complete graph K_n in two colors. In

108

Theorem 10.1 we found G_n such that $G_n = \overline{G}_n$. In a sense, we tiled K_n by two copies of G_n.

Let k be a positive integer. We can color edges of K_n in k colors and pose a question: *find all n such that K_n can be tiled by k copies of a tile G_n.*

Exercise 10.6. Prove that if K_n is tileable by m copies of a tile G_n, then

$$m \text{ is a divisor of } \binom{n}{2} \qquad (*)$$

where $\binom{n}{2} = \frac{n(n-1)}{2}$.

It is relatively easy to show that *the condition (*) is also sufficient for any odd k.* For example, let us prove it for $m = 5$ and call it **Example 10.4**.

Solution of Example 10.4. Due to Exercise 10.6, we must have

$$n \equiv 0 \pmod{5}$$

$$\text{or}$$

$$n \equiv 1 \pmod{5}$$

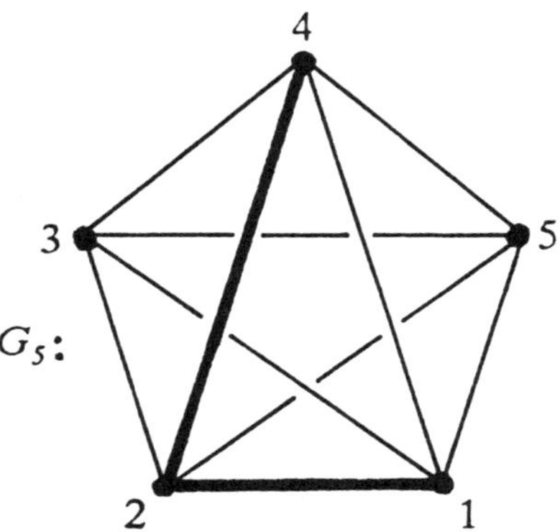

Figure 10.11

For $n = 1$ the result is obviously true. For $n = 5$ we can easily solve it: in Figure 10.11 bold lines show the edges of one of the five tiles G_5 (other tiles can be obtained by rotations of this one about the center of the pentagon).

Now, similarly to the proof of Theorem 10.1, we need to show how to *sew* solutions for larger values of n.

For $n = 10$ we take two copies of G_5 and *sew* them together to get G_{10} as shown in Figure 10.12. We do it, of course, for each of the five colors.

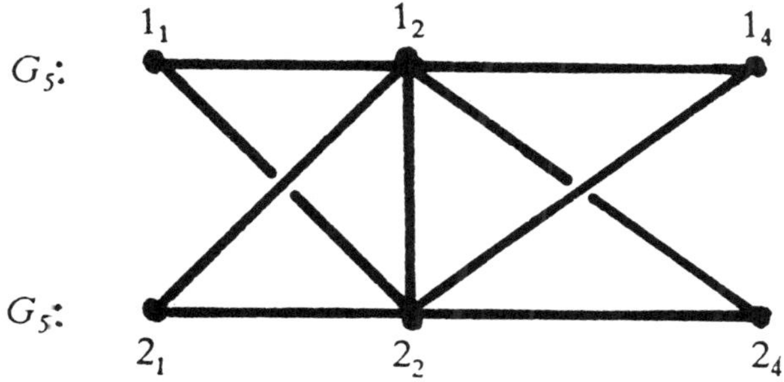

Figure 10.12

Let $n = 5m$, where m is an integer greater than 1. We construct m copies of G_5 and *sew* every two of them together as shown in Figure 10.12 to get G_{5m}. We do it for each of the five colors. As a result, K_{5m} is tiled by five copies of the tile G_{5m}.

Now let us solve the case $n = 5m + 1$.

For $m = 0$ the solution G_1 is trivial (Figure 10.13). When $m = 1$ we expand our *sew* operation to connect G_1 and G_5. The result is G_6 shown in Figure 10.14.

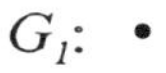

G_1: •

Figure 10.13

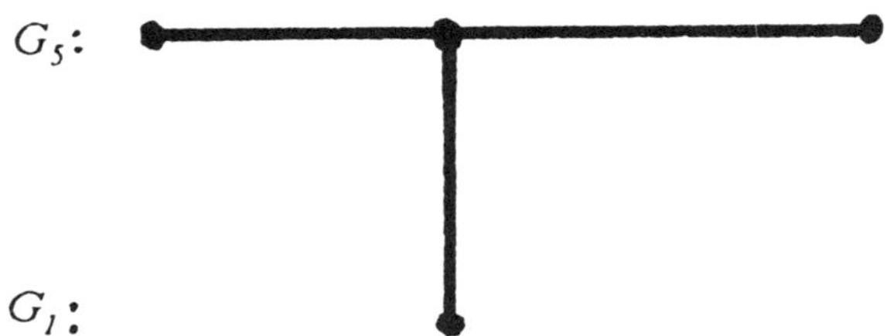

Figure 10.14

The tile graph G_{5m+1} ($k \geq 1$) is constructed from m copies of G_5 and one copy of G_1 by *sewing* every two of them together. It is not difficult to show (please do) that K_{5m+1} can be tiled by five copies of the tile G_{5m+1}.

This construction allows a straightforward generalization.

Exercise 10.7. For every odd divisor m of $\binom{n}{2}$ construct a tile graph G_n such that K_n can be tiled by m copies of G_n.

This problem has been solved completely by F. Harary, R.W. Robinson, and N. Wormald in [HRW]. They showed that the condition (*) is always sufficient.

In Chapter 1 we discussed tiling of rectangular boards by polyominal tiles. This problem can be translated into the language of graph theory. Indeed, we can represent every unit square of the board by a vertex and connect two vertices by an edge if and only if the corresponding squares are adjacent (Figure 10.15).

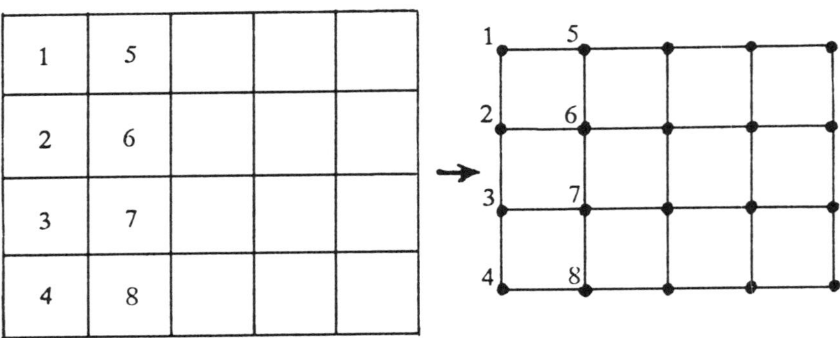

Figure 10.15

And, of course, a polyominal tile will become a connected graph.

Tiling a rectangle is essentially no different from cutting it into tiles. But cutting the board along the checker lines translates into cutting edges of the graph (Figure 10.16). For simplicity *cutting a graph* G would stand for cutting (removing) some of the edges of G. Now we can translate the whole problem.

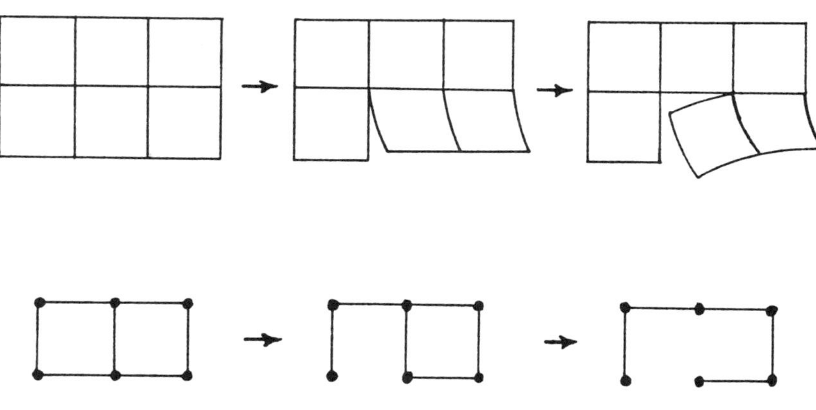

Figure 10.16

Given two connected graphs G and T. Cut G into components each isomorphic to T, or simply into tiles T.

Exercise 10.8. If a graph G can be cut into components T, then

$$|T| \text{ is a divisor of } |G|.$$

Exercise 10.9. Let P_3 be a tromino graph (Figure 10.17). Then K_n can be cut into tiles P_3 if and only if n if a multiple of 3.

Figure 10.17

Exercise 10.10. A complete bigraph $K_{m,n}$ ($m \geq n$) can be cut into tiles P_3 if and only if $m + n$ is a multiple of 3 and m,n satisfy the isosceles triangle inequality:

$$m \leq n + n$$

In our discussions above we could not differentiate between linear tromino and L-tromino. They both are represented by the same tile graph P_3. We can resolve this situation by coloring edges of a graph to be tiled in two colors.

Let G be the graph of a rectangular board (Figure 10.18). Let us color all horizontal edges in one color (bold segments in Figure 10.18), and vertical edges in the other color. Now we can differentiate between the two types of trominoes. The linear tromino produces a tile P_3 with both edges of the same color, and L-tromino corresponds to a P_3 with one edge of each color.

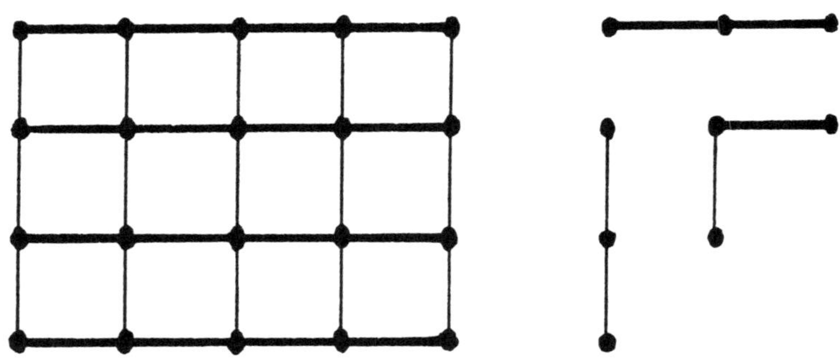

Figure 10.18

Thus, the problem of tiling with L-tromino can be translated into the language of graphs as follows.

Edges of a connected graph G are colored in two colors. Cut G into tiles P_3 containing one edge of each color.

The problem of packing a parallelepiped can also be translated into the language of graph theory. We can represent every unit cube of the parallelepiped by a vertex and connect two vertices by an edge if and only if the corresponding unit cubes are adjacent (Figure 10.19).

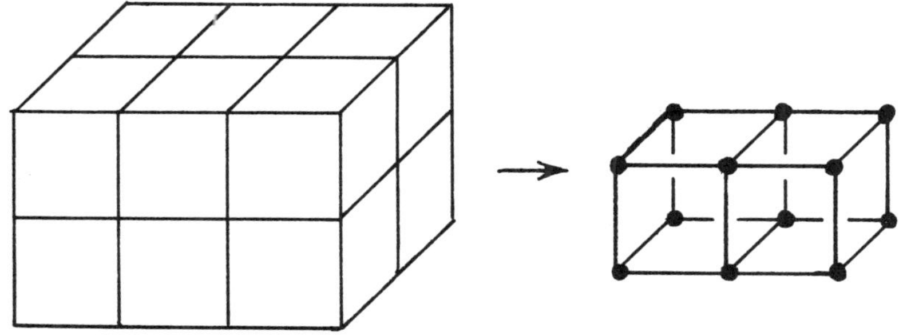

Figure 10.19

If we wish to differentiate between various 3-dimensional tiles, we may color the edges of the graph in three colors in accordance with the coordinate axes.

Solutions of Exercises

10.1. See the solution of Exercise 9.3 of the previous section.

10.2. Let G be a graph of order n such that $G = \overline{G}$. Then the number of edges in K_n must be even, i.e., $\frac{n(n-1)}{2}$ is even. But since one of the numbers n, $n-1$ is odd, the other of them must be divisible by 4, which proves the required statement.

10.3. See:

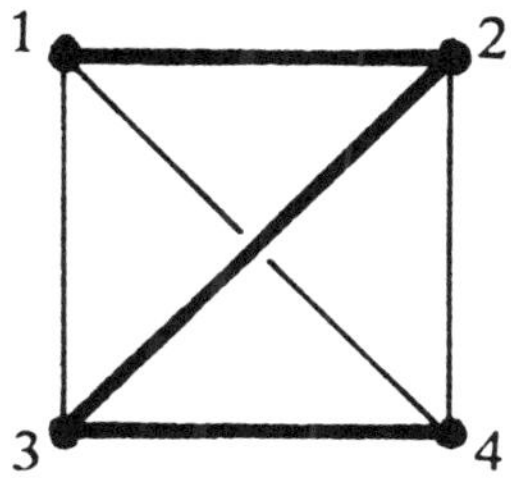

Figure 10.20

10.4. See Figure 10.6.

10.5. Yes, in fact for any $m > 1$ we can construct an alternative $G'_{4m} \neq G_{4m}$ and for any $m \geq 1$ an alternative $G'_{4m+1} \neq G_{4m+1}$. We can *sew* the basic pieces differently (Figures 10.21 and 10.22):

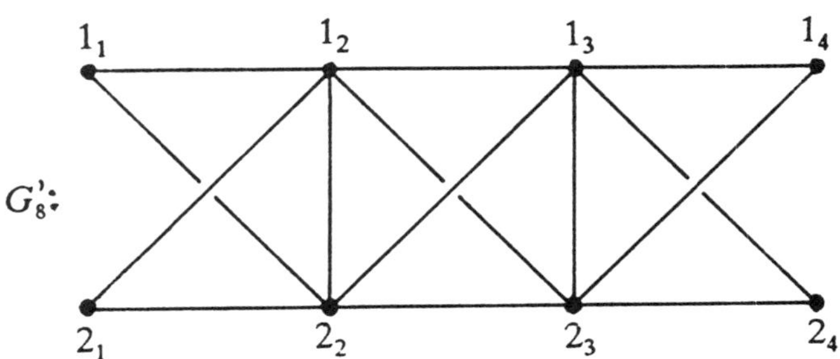

Figure 10.21

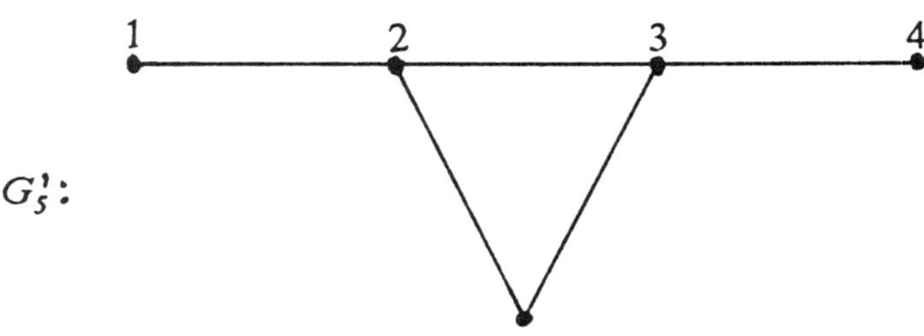

Figure 10.22

10.6. Obviously, m must be a divisor of the number of edges of K_n, which is exactly $\binom{n}{2}$.

10.7. The construction is similar to the one in Example 10.4.

10.8. When the graph G is partitioned into tiles T, the $|G|$ vertices are thus partitioned into groups of $|T|$ vertices, therefore $|T|$ is a divisor of $|G|$.

10.9. Due to Exercise 10.8, n must be a multiple of 3, i.e., $n = 3k$. If $n = 3k$ then we can partition n vertices into groups of three, save in every group two edges and cut all other edges.

10.10. Let V_1 and V_2 be the subsets of m and n non-adjacent vertices respectively that form the graph $K_{m,n}$ ($m \geq n$). Assume that $K_{m,n}$ is cut into tiles P_3. Then at least one vertex of each tile must belong to V_2 (remember, there are no edges between any two vertices of V_1. Therefore, $m \leq n + n$. In addition, of course, $m+n$ is a multiple of 3 (Exercise 10.8).

Conversely, assume that positive integers m,n satisfy the inequality $n \leq m \leq n + n$ and $m + n$ is a multiple of 3. Let us take a look at numbers $2m - n$ and $2n - m$. They are positive. They both are divisible by 3 because $2m - n = 3m - (m + n)$ and $2n - m = 3n - (m + n)$. Now construction is clear. We line up $\frac{2m-n}{3}$ copies of "V" followed by $\frac{2n-m}{3}$ copies of "Λ" (Figure 10.23).

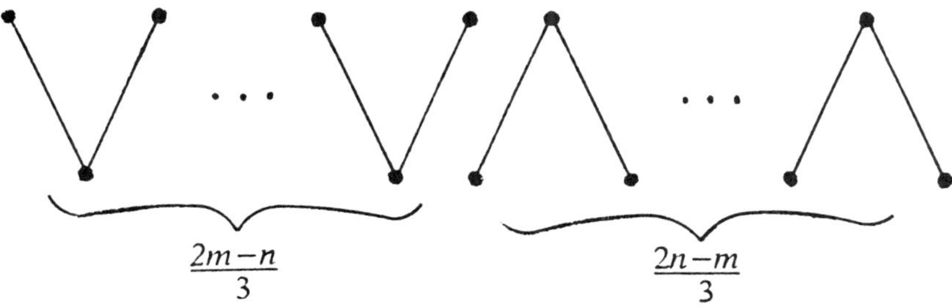

$$\frac{2m-n}{3} \qquad\qquad \frac{2n-m}{3}$$

Figure 10.23

In the top row of Figure 10.23 we have

$$2\frac{2m-n}{3} + \frac{2n-m}{3} = m$$

vertices. In the lower row we have

$$2\frac{2n-m}{3} + \frac{2m-n}{3} = n$$

vertices. Therefore, Figure 10.23 shows exactly how to cut $K_{m,n}$ into P_3 tiles.

11. PLANARITY

How would you draw a complete graph K_4 in the plane?
Probably like Figure 11.1.

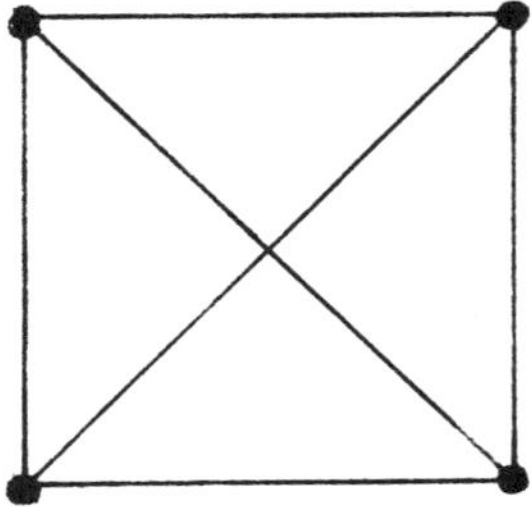

Figure 11.1

In Figure 11.1 two edges of K_4 intersect. Can we avoid it,
i.e., draw K_4 in the plane without any intersections
of edges? Yes. Remember we can think of a graph as a set of
pins, some of which are connected by rubber bands. So
we can pick up one of the diagonal rubber bands and pull it out
(Figure 11.2). We can also redraw K_4 as in Figure 11.3.

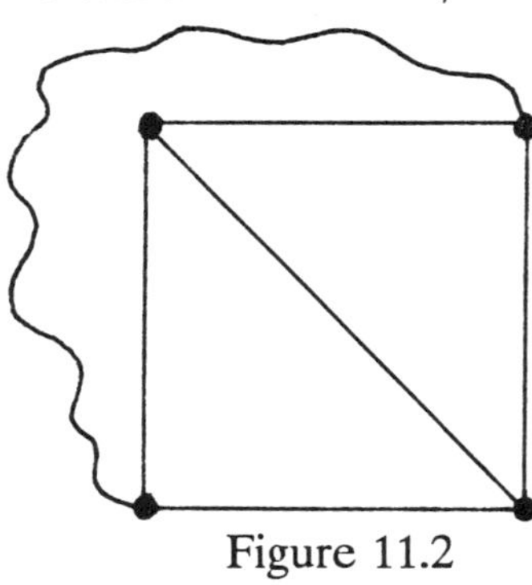

Figure 11.2

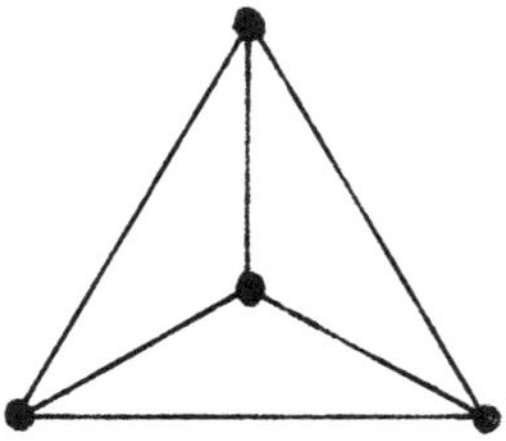

Figure 11.3

A graph that can be drawn in the plane without any intersections of edges is called *planar.* Otherwise a graph is called *non-planar.* If a planar graph *is* already drawn in the plane without intersections of edges, it is called a *plane graph.*

A *region* of a plane graph G is a maximal part of the plane for which any two points may be joined by a curve that does not intersect edges or vertices of G. If we think of a graph G on the plane as a sheet of paper, the regions of G would be the pieces into which the plane would be divided if we were to cut it along all the edges of G.

Every region of a plane graph G has a *boundary* that consists of vertices and edges of G. Every plane graph G also has an unbounded region called the *exterior region* of G.

Planar graphs are associated with the study of polyhedra. In fact, the vertices and edges of any convex polyhedron P (the definition of a convex polyhedron see in Section 13, p. 131) form a graph $G(P)$ called the *skeleton* of P.

This graph $G(P)$ is connected and planar for every convex polyhedron P (can you prove that?). Therefore everything we prove for planar graphs, will be applicable to the skeletons of polyhedra.

Every vertex of $G(P)$ has degree at least three.

The faces of P correspond to the regions of $G(P)$.

Exercise 11.1. Prove the following formula for the graph $G(P)$ of a polyhedron P:

$$3F_3 + 4F_4 + ... + nF_n = 2q \qquad (1)$$

where q is the number of edges of P, F_3 is the number of triangular faces of P, F_4 is the number of quadrangular faces ,..., F_n is the number of n-gonal faces (where n is the greatest number of edges bounding a face of P).

Exercise 11.1 has the following very important corollary. Can you prove it?

Corollary 11.1. *For any planar graph G*

$$3F_3 + 4F_4 + ... + nF_n \leq 2q \qquad (1')$$

where q is the number of edges of G; F_3, F_4, ..., F_n are the numbers of triangular, quadrangular, ..., n-gonal faces of G respectively.

Let us prove a celebrated formula discovered by one of the greatest mathematicians of all times, Leonard Euler (1707-1783).

Theorem 11.1. (***Euler Formula for Graphs***) *If G is a connected plane graph with p vertices, q edges, and r regions, then*

$$p + r = q + 2 \qquad (2)$$

Proof. We will prove the equality (2) by induction on q

For $q = 1$ the result is obvious since in this case $p = 2$ and $r = 1$.

Assume the equality (2) is true for all connected plane graphs with fewer than q edges.

Let G be a connected plane graph with q edges (and p vertices and r regions). We will consider two cases.

Case 1. If G has no cycles, then G contains a vertex of degree one. Indeed from any vertex v_1 of G we can travel along an edge to an adjacent vertex v_2. If v_2 is not of degree one, we continue our travel to v_3, etc. Since we can never come again to any of the vertices we previously visited (no cycles, remember?), and the total number of vertices is finite, we must end up in a vertex v_n of degree one. Now we remove v_n together with the edge incident to v_n. We get a graph G' with $q - 1$ edges, $p - 1$ vertices and r regions. Since by inductive assumption the equality (2) is true for G', we get

$$(p - 1) + r = (q - 1) + 2$$

By adding one to both sides of the last equality, we prove the required equality (2) for the graph G.

Case 2. If G has a cycle, we remove one edge of the cycle. We get the graph G' that is connected (do you see why?) and plane, and has $q - 1$ edges, p vertices, and $r - 1$ regions. By inductive assumption the equality (2) is true for G', i.e.,

$$p + (r - 1) = (q - 1) + 2$$

which implies the equality (2) for G.

As we mentioned before (pp. 105-106), the results from the theory of planar graphs can be applied to polyhedra. Thus we get the following important and beautiful corollary.

Corollary 11.2. (*Euler Formula for Polyhedra*) If $V, E,$ and F are the numbers of vertices, edges, and faces of a convex polyhedron, then

$$V + F = E + 2 \tag{2'}$$

Exercise 11.2. Prove that in any planar graph with p vertices and q edges

$$q \leq 3(p - 2) \tag{3}$$

Exercise 11.3. Prove that in any planar graph without triangles, with p vertices and q edges

$$q \leq 2(p - 2) \tag{4}$$

Exercise 11.4. If a graph G contains a non-planar subgraph, then G is non-planar.

Exercise 11.5. Prove that a complete graph K_n is planar if and only if $n < 5$.

Exercise 11.6. Prove that a complete bigraph $K_{m,n}$ is planar if and only if at least one of the positive integers m, n is less than three.

An *elementary subdivision* of a graph G (with at least one edge) is a graph obtained from G by inserting an additional vertex in an inside point of some edge of G. Two graphs G_1 and G_2 are said to be *homeomorphic* if each of them "comes from the same ancestor" G, i.e., each of them can be obtained from the same graph G by a succession of elementary subdivisions.

Exercise 11.7. Prove that the graphs G_1 and G_2 in Figure 11.4 are *homeomorphic*.

120

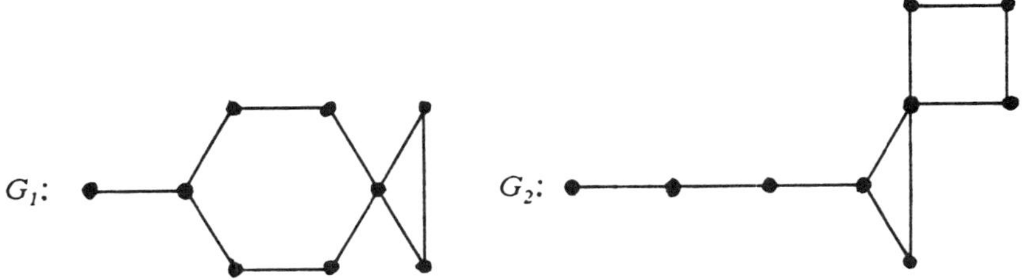

Figure 11.4

Now we are ready for a brilliant criterion of planarity that was discovered by the famous Polish mathematician Kazimir Kuratowski in 1930 ([Ku]).

Theorem 11.2. *(Kuratowski Theorem)* *A graph is planar if and only if it contains no subgraph homeomorphic to K_5 or $K_{3,3}$.*

Exercise 11.8. Prove that the Petersen graph in Figure 11.5 is nonplanar.

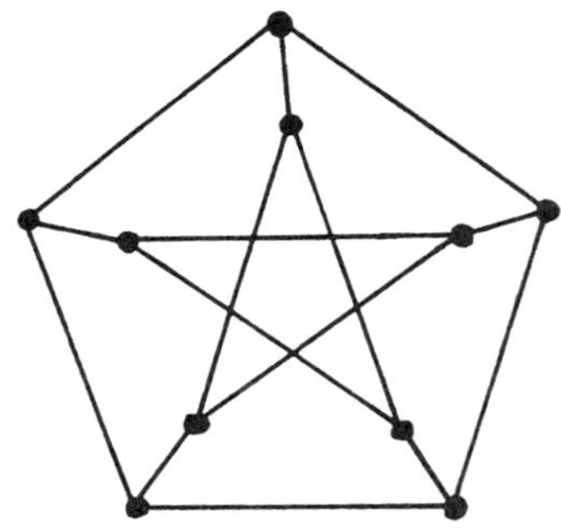

Figure 11.5

You can prove (do, and let us call it **Exercise 11.9**) that any graph can be drawn in the plane so that every edge is a straight line segment.

The following result though is not at all obvious.

Theorem 11.3. (*Wagner* [W], Fáry [F]) *A planar graph can be drawn in the plane without intersections of edges and so that every edge is a straight line segment.*

We divided all connected graphs into planar and non-planar. Are all nonplanar graphs alike? No, and we have a couple of invariants for you.

There are many ways to draw a graph in the plane, and, of course, the number of crossings of its edges may vary. The **crossing number** $v(G)$ of a graph G is the minimum number of crossings of its edges among all the drawings of G in the plane.

The **rectilinear crossing number** $\overline{v}(G)$ of a graph G is the minimum number of crossings of its edges taken over all those drawings of G in the plane in which every edge is a straight line segment.

Of course,

$$v(G) \leq \overline{v}(G)$$

But do we really need two crossing numbers? For planar graphs $v(G) = \overline{v}(G) = 0$ (Theorem 11.3). Moreover, Richard Guy proved in [Gu] that for any graph G with seven points or less $v(G) = \overline{v}(G)$. Yet, in the same paper he showed that the complete graph K_8 delivers a counter example:

$$v(K_8) = 18$$

where as

$$\overline{v}(K_8) = 19$$

So not everything in the plane can be reduced to drawing straight line segments.

What about the *three-dimensional space* R^3? Can every graph be drawn in R^3 without crossing of its edges and so that every edge is a straight line segment?

Theorem 11.4. *Yes! Every graph can be drawn in R^3 so that every edge is a straight line segment and no crossing occurs.*

Proof. Place the vertices of the given graph G in the *points of general position*, i.e., no four points in a plane, and no three points on a line. Now connect the adjacent points of G by line segments. No crossing will occur, because otherwise four vertices would lie in a plane.

We are done. Are we?

Does a set of n points of general position exist for any positive integer n? Yes, and we can construct it by induction.

The first two distinct points we can pick arbitrarily, the third point we chose so that not all three points are on a line.

Assume that n points of general position are chosen ($n \geq$ 3). Through every three points we draw a plane. We get finitely many planes (namely $m = \binom{n}{3}$). They do not exhaust the space so we can pick the $(n+1)$st point outside of the drawn planes.

We are done. Are we?

Well, yes, if you believe that m planes (m being an integer) do not exhaust the space. If you do not believe it, pick a plane P not parallel to (and not one that coincides with) any of the m planes. The m planes will intersect P along m straight lines, which do not exhaust P, so we can pick the $(n+1)$st point on P outside of the m lines.

We are done. Are we?

Well, yes, if you believe that m straight lines do not exhaust a plane P. If you do not believe it, pick a straight line L in P which is not parallel to (or coincides with) any of the m lines. The m lines will intersect L in m points, which do not exhaust L, so we can pick the $(n + 1)$st point in L distinct from the m points.

We are done. Are we?

Solution of Exercises.

11.1. There are two ways to add up the number T of edges on the boundaries of all faces. Triangular faces are bound by three edges, quadrangular by four, ..., n-gular faces are bounded by n edges, therefore

$$T = 3F_3 + 4F_4 + ... + nF_n$$

On the other hand, T counts every edge exactly twice because every edge e is counted on the boundary of two faces that are separated by e (or twice on the single face that is adjacent to it on both sides, as Figure 11.4). Therefore, $T = 2q$.

11.2. The inequality (1') of corollary 11.1 implies

$$3(F_3 + F_4 + \dots + F_n) \leq 2q$$

i.e.,

$$3r \leq 2q \tag{5}$$

where r is the total number of regions of G.

Combining the Euler formula (2) with (5), we get

$$(p + \tfrac{2}{3}q) \geq q + 2$$

i.e.,

$$q \leq 3(p-2)$$

11.3. Since $F_3 = 0$, the inequality (1') of Corollary 11.1 implies

$$4(F_4 + \dots + F_n) \leq 2q$$

i.e., $$\tag{6}$$

$$4r \leq 2q$$

where r is the total number of regions of G. Now the Euler formula (2) and (6) combine to produce the required inequality:

$$p + \tfrac{1}{2}q \geq q + 2$$

i.e.,

$$q \leq 2(p - 2)$$

11.4. Argue by contradiction.

11.5. We already know that K_n is planar for $n < 5$ (Figure 11.1 shows it for $n = 4$). Let us take a look at K_5. It has $p = 5$ vertices and $q = 10$ edges. Assume that K_5 is planar, then by Exercise 11.2 it would satisfy the inequality $q \leq 3(p-2)$, i.e., $10 \leq 9$ which is absurd. Therefore K_5 is not planar.

Finally, every graph K_n for $n \geq 5$ contains K_5 as a subgraph. Therefore, K_n is not planar for $n \geq 5$.

11.6. Figure 11.6 shows that $K_{m,2}$ is planar for any positive integer m.

Therefore any subgraph of $K_{m,2}$ is planar as well, i.e., a complete bigraph $K_{m,n}$ is planar if at least one of the positive integers m,n is less than three.

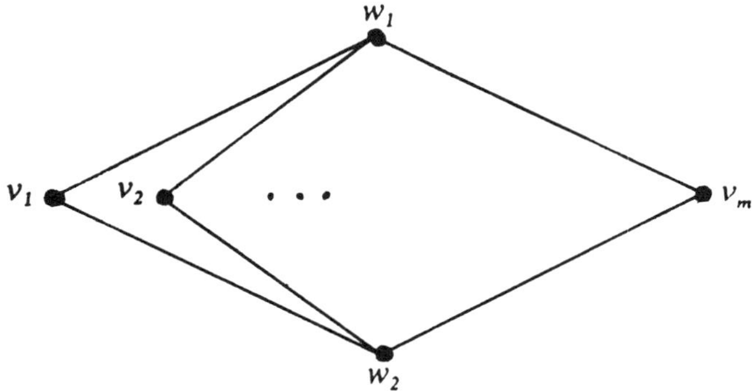

Figure 11.6

Let us take a look at $K_{3,3}$. It has $p = 6$ vertices, $q = 9$ edges, and has no triangles (Theorem 9.2). Therefore, by Exercise 11.3' $q \leq 2(p-2)$, i.e., $9 \leq 8$, which is absurd. Therefore, $K_{3,3}$ is not planar.

Every graph $K_{m,n}$ for $m \geq 3$, $n \geq 3$ contains $K_{3,3}$ as a subgraph, and therefore is not planar.

11.7. Both graphs G_1 and G_2 can be obtained from the graph G (Figure 11.7) by a succession of elementary subdivisions.

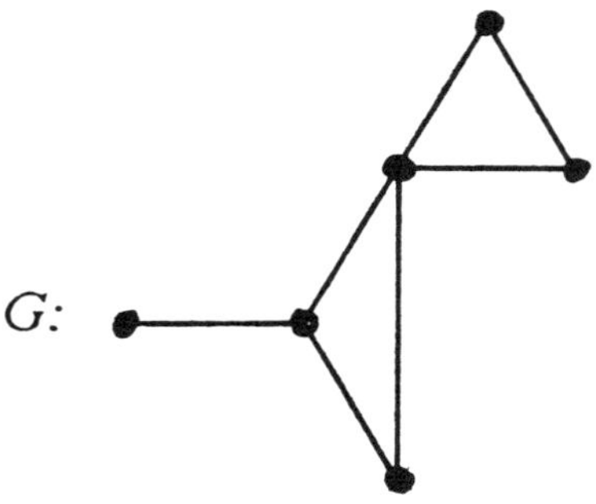

G:

Figure 11.7

11.8. See:

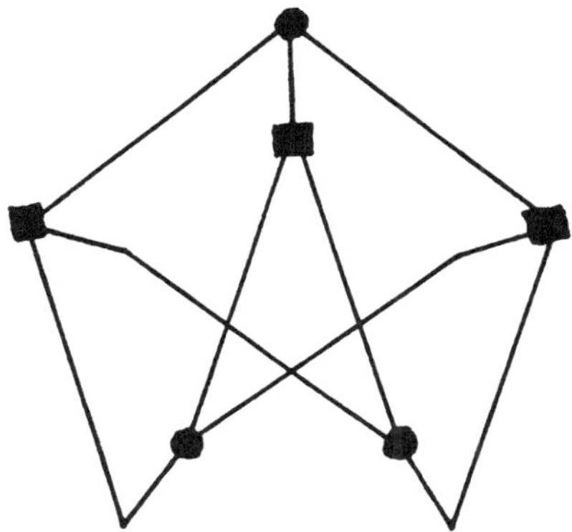

Figure 11.8

12. THE INTERSECTION INDEX AND THE JORDAN THEOREM

In the previous section we talked about *regions* that we obtain when a graph is drawn in the plane without intersections. Is it clear what a *region* is? Perhaps you would answer *yes*. But allow us to shake up your confidence in the clarity of this notion.

Let us consider a very simple example: a complete graph K_3, that is the contour of a triangle (Figure 12.1). Is it evident that the graph in Figure 12.1 divides the plane into two regions? In other words, is it true that every closed loop that does not intersect itself defines two regions, interior and exterior?

It appears "visually obvious" that a curve C can not be a common boundary of more than two regions in the plane, such that all regions border on c along the entire C. Intuition, however, tricks us.

126

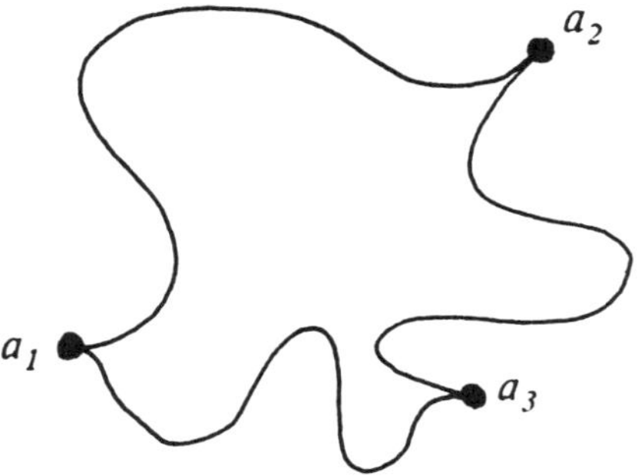

Figure 12.1

Example 12.1. *There is a curve in the plane, that is a common boundary of three regions.*

Such curves known as the "Lakes of Wada" were discovered by the Japanese mathematician, Yoneyama in 1917 ([Y]).

Proof. Assume that a portion of land, surrounded by the sea, contains two lakes: a warm lake and a cold lake. In order to provide the land with water we build canals.

On the first day we build a canal (Figure 12.2)) that delivers the warm lake water so that it is available at a distance not exceeding 1 from every point of land (and this canal is neither connected to the sea nor to the cold lake!).

Figure 12.2

On the second day we build a canal that delivers the cold lake water so that it is available at a distance not exceeding 1 from every point of the remaining land (and this canal may not be connected to the sea, the warm lake, or to the previously built canal).

On the third day we build a canal that delivers the sea water so that it is available at a distance not exceeding 1 from every point of the remaining land (and, of course, this canal may not be connected to the lakes or to the previously built canals).

During the next three days we extend the three canals further in such a way that the warm lake water, the cold lake water, and the sea water are available from any point of the remaining land at a distance not exceeding 1/2.

During the following three days we increase the density of the canals so that every type of water is available from every point of the remaining land at a distance not exceeding 1/4, etc. Please note that after each day, the remaining land is a connected piece; thus, the next day we can build on it.

In the limit we will get a system of the warm lake, the cold lake, and the sea waters that do not mix anywhere. What is left of the land will be a *curve*, such that the warm lake, the cold lake, and the sea waters will be however close from any point of this curve c. In other words, all three regions--the warm lake with its canal, the cold water lake with its canal, and the sea with its canal--will border on c along the entire curve.

Certainly the curve c described in this example is very complicated. But this example can sow a doubt in your mind: perhaps a non-self-intersecting closed curve hides similar surprises? Of course, you may say that each graph contains only a finite number of edges, and it is possible to limit ourselves to broken lines. Indeed, you probably think that for broken lines all facts are simple and visually clear, are they not? Let us, however, take a look at the following example. In Figure 12.3 we have a closed broken line (you can verify that by following the line with a pencil). But is it really "clear" that this closed line divides the plane into two regions? (Is it clear whether a policeman A and a bandit B are in the same region?) And why is it clear? Is it an axiom, a theorem, or a non-demonstrable fact?

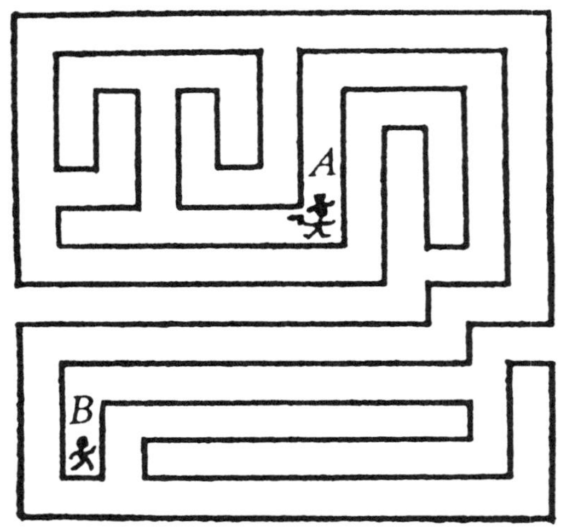

Figure 12.3

In the last century a well-known French mathematician Camille Jordan proved the following proposition. In his theorem, Jordan considers a simple closed curve that is a continuous one-to-one image of a circle.

Theorem 12.1. *(Jordan Theorem) Each simple closed curve C in the plane divides the plane into two regions.*

Every two points situated in one region can be joined by a broken line not intersecting C, whereas each broken line joining two points in different regions intersects C.

In this section we are going to prove this theorem for an arbitrary broken contour C (like in Figure 12.3), that is, for a closed broken non-self-intersecting line. For this purpose we need the notion of the intersection index.

Let S and T be two segments in the plane in general position, i.e., none of them passes through an end point of the other. If they intersect, then we write $I(S,T)=1$; otherwise $I(S,T)=0$. The number $I(S,T)$ is said to be the **intersection index** of the segments S and T.

For convenience, any broken line in the plane (maybe with self-intersections) will be called a *chain*. A closed chain (that is a closed broken line) will be called a *cycle*. Let X be a chain formed by segments $S_1, ..., S_m$, and Y be a chain formed by segments $T_1, ..., T_n$. Suppose that the chains X and Y are in a general position, that is *every* segment S_i is in general position with every segment T_j. If the sum $\Sigma\, I(S_i, T_j)$ (over all $i = 1, ..., m; j = 1, ... n$) is *even*, then we write $I(X,Y) = 0$; otherwise $I(X,Y) = 1$. We say that $I(X,Y)$ is the **intersection index of the chains** X and Y (more precisely, the intersection index modulo 2).

Exercise 12.1. Let X and Y be two cycles and L be a line that is not parallel to any line joining a vertex of X with a vertex of Y. We move the cycle Y continuously in the direction of the line L. Prove that the intersection index of the cycle X with the moving cycle Y does not change, that is $I(X,Y_1) = I(X,Y_2)$ for any two positions Y_1 and Y_2 of Y that are in general position with X.

Exercise 12.2. Prove that the intersection index of every two cycles X, Y in general position is equal to 0.

Example 12.2. *Using the intersection index, we can prove (in another manner than in Section 11) that the graph K_5 is not planar.*

Indeed, we draw all the edges of the graph K_5 (with intersections) as broken lines in general position with each other, and denote by I the sum of the numbers of intersection points over all pairs of *non-adjacent* edges, taken modulo 2. Suppose that we changed the position of only one edge, say $a_1 a_2$ (Figure 12.4). Let X_0 be the old position of the edge $a_1 a_2$, and X_1 be its new position. Then the union of these chains is a cycle; we denote it by $X_0 + X_1$. Further, the union of the edges $a_3 a_4$, $a_4 a_5$, $a_5 a_3$ (that is, the edges non-adjacent with $a_1 a_2$) is a cycle too. We denote it by Y. We know (cf. Exercise 12.2) that $I(X_0 + X_1, Y) = 0$, that is $I(X_0, Y = I(X_1, Y)$. This means that changing the position of *one* edge does not change the value of I.

But changing one by one the positions of edges we can pass from one drawing of the graph K_5 to any other. So, the value of I is *the same* for all drawings of the graph K_5. But for the drawing in Figure 12.5 we have $I = 1$. Consequently, $I = 1$ for

130

every drawing of the graph. This means that it is impossible to draw the graph K_5 without intersecting of edges, that is, this graph is non-planar.

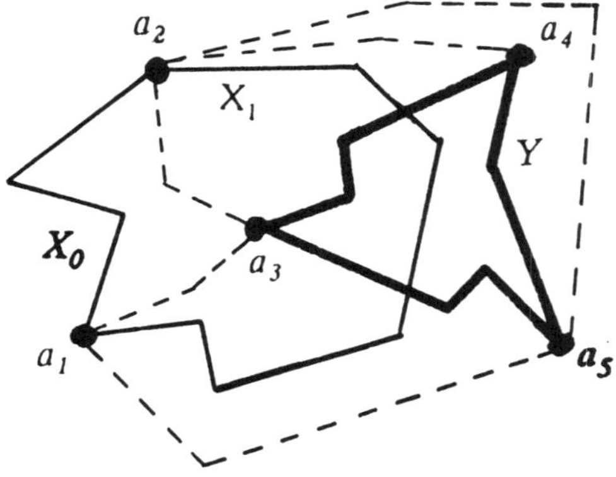

Figure 12.4

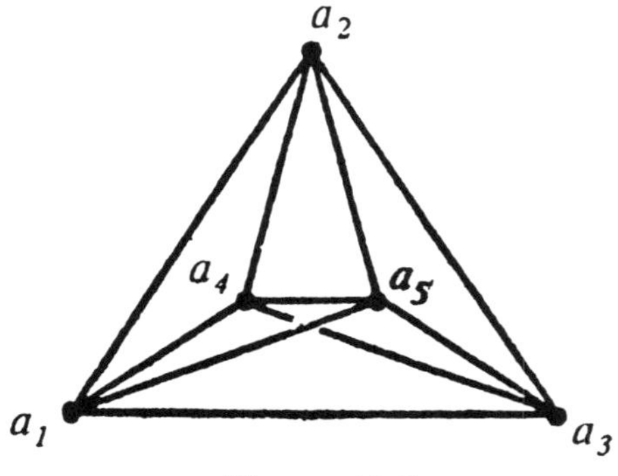

Figure 12.5

Exercise 12.3. Using the intersection index, prove that the graph $K_{3,3}$ is non-planar.

Exercise 12.4. Prove that there exist two cycles X, Y on the torus, such that $I(X,Y) = 1$. Does this mean that the graph K_5 (or $K_{3,3}$) can be embedded in the torus (without intersections of edges)?

Now let us return to the Jordan Theorem (Theorem 12.1) and present a proof of this theorem for a closed non-intersecting broken line C.

Let S_1, S_2, ..., S_n be a consecutive listing of all segments of C. We take points p, p' that are symmetric with respect to S_1 (Figure 12.6). We construct the segment through p parallel to S_1 until intersection with the bisector of the angle between S_1 and S_2.

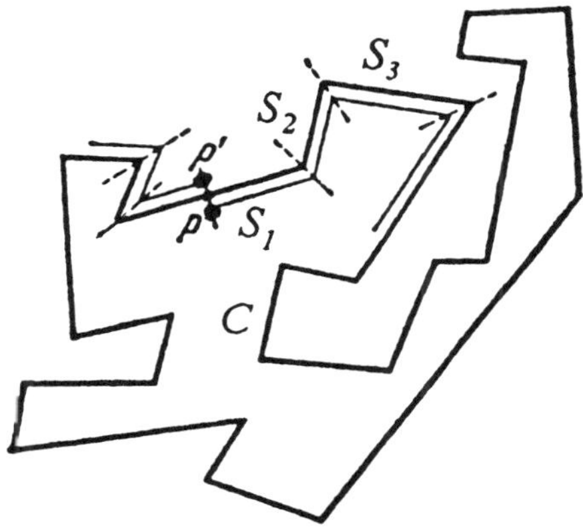

Figure 12.6

From the intersection point we carry out the segment parallel to S_2 till intersection with the bisector of the angle between S_2 and S_3, etc. We end up with a broken line M, such that its segments are all on the same distance from the corresponding segments of the initial broken line C. If $|pp'|$ is small enough, then C and M don't intersect. Consequently, going around C the broken line M must return either to p, or to p'. *But M cannot come to p', because otherwise the union of M and the segment $[p, p']$ is a cycle Y (Figure 12.7), such that $I(C, Y) = 1$, contradicting the result of* Exercise 12.2. Thus, M is a closed broken line one time going around C and returning to p.

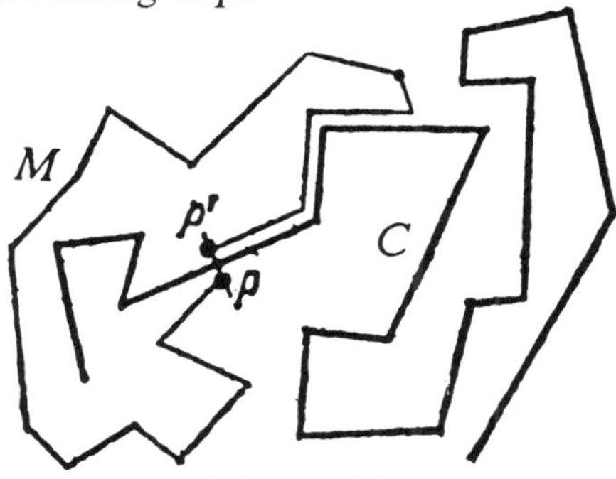

Figure 12.7

Similarly, we construct the broken line M' starting from p', going around C and returning to p'.

Let now q be an arbitrary point that does not belong to C. Then, without intersecting C, it is possible to join it with either p or p'. Indeed, let us draw a ray emanating from q that intersects M and M'. Now we can go along this ray from q till *the first* point of intersection with either M or M' and then along this broken line till either p or p'.

It can be easily shown that if Y and Z are two broken lines non-intersecting with C that both start in q and end at one of the points p or p', then they must end up at *the same* point (otherwise, the union X of Y,Z and the segment $[p,p']$ would be a cycle with $I(X,L) = 1$, contradicting the result of Exercise 12.2).

Let us now denote by U the set of all points connectable with p and by V the set of all points connectable with p' without intersecting C. Then U and V are the two regions mentioned in the statement of the Jordan Theorem. Indeed, if points q_1, q_2 *belong to the same region (say, U)*, then there exists a broken line joining q_1 and q_2 without intersecting C. But if the points q_1 and q_2 belong to different regions, then it is impossible to join them without intersecting C (otherwise we would obtain, as above, a cycle X with $I(X,C) = 1$). We are done.

Let us note that all "far away" points of the plane are situated in the same region; this region is said to be the *exterior* region of C. The other region is called the *interior*. This means that the interior region is bounded, whereas the exterior region is unbounded.

Solutions of Exercises

12.1. The intersection index $I(X,Y)$ could change only at the moments when a vertex of one cycle coincides with an interior point of a segment of the other cycle. (We note that a vertex of X cannot coincide with a vertex of the moving cycle Y by virtue of the choice of L.) However, in those moments the parity of the number of intersection points does not change (Figure 12.8). Thus, the intersection index does not change.

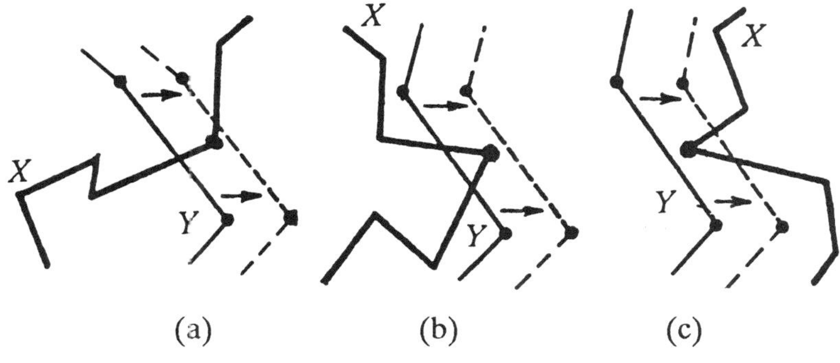

(a)   (b)   (c)

Figure 12.8

12.2. Let us move the cycle Y, as in Exercise 12.1 from an initial position Y_0 to a position Y_1 where Y does not intersect X (Figure 12.9). Then in the position Y_1 the intersection index is equal to 0. Hence, by virtue of Exercise 12.1, for the initial position $Y = Y_0$ we have $I(X,Y_0) = I(X,Y_1) = 0$.

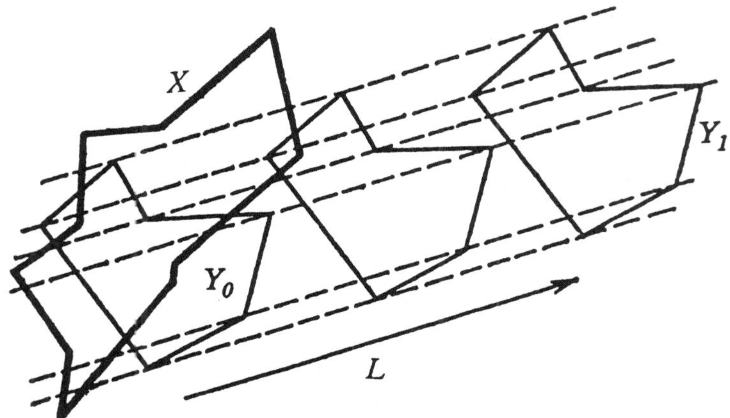

Figure 12.9

12.3. We draw all edges as broken lines in general positions (perhaps with intersections) and denote by I the sum modulo 2 of the numbers of intersection points on all pairs of *non-adjacent* edges. Suppose that we changed the position of only one edge, say a_1b_1. The initial position of the edge we denote by X_0 and the new position by X_1 (Figure 12.10). The union of those edges is a

134

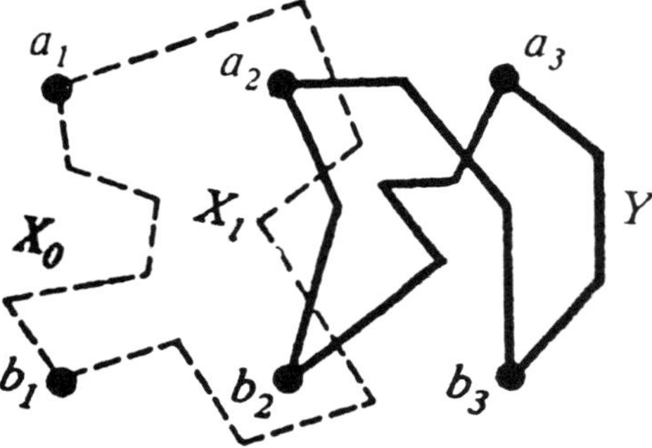

Figure 12.10

cycle $X_0 + X_1$. Four edges a_2b_2, a_2b_3, a_3b_2, a_3b_3 are non-adjacent with a_1b_1. They form a cycle too; we denote it by Y. By virtue of Exercise 12.2, we have $I(X_0 + Y_1, Y) = 0$, that is $I(X_0, Y) = I(X_1, Y)$. This means that changing the position of *one* edge does not change the value of I. By changing one by one the positions of edges we conclude that the value of I is *the same* for all drawings of the graph $K_{3,3}$. But for the drawing in Figure 12.11 we have $I = 1$. Consequently, $I = 1$ for *every* drawing of the graph. This means that it is impossible to draw the graph $K_{3,3}$ without intersections of edges, that is, the graph $K_{3,3}$ is non-planar.

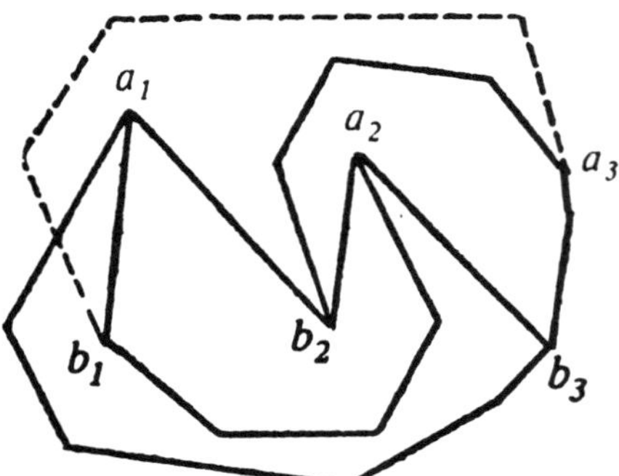

Figure 12.11

12.4. In Figure 12.12 two cycles X and Y are drawn on the torus that have only one intersection point C (see page 50 for the definition of a torus). Thus, $I(X,Y) = 1$. This means that the reasoning of Example 12.2 and the solution of Exercise 12.3 fail in the case of a torus. This gives hope that the graphs K_5 and $K_{3,3}$

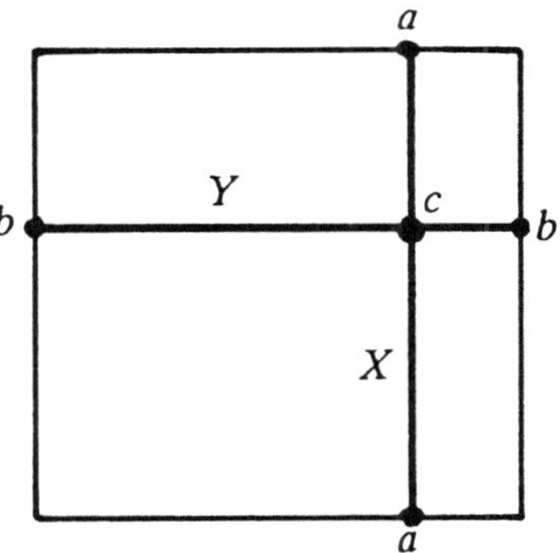

Figure 12.12

can be embedded in the torus (without intersections of edges). Indeed, an embedding of K_5 is drawn in Figure 12.13, and an embedding of $K_{3,3}$ is shown in Figure 12.14 (see Figure 12.11). But it should be noted that there exist graphs that cannot be embedded in the torus (without intersections of their edges).

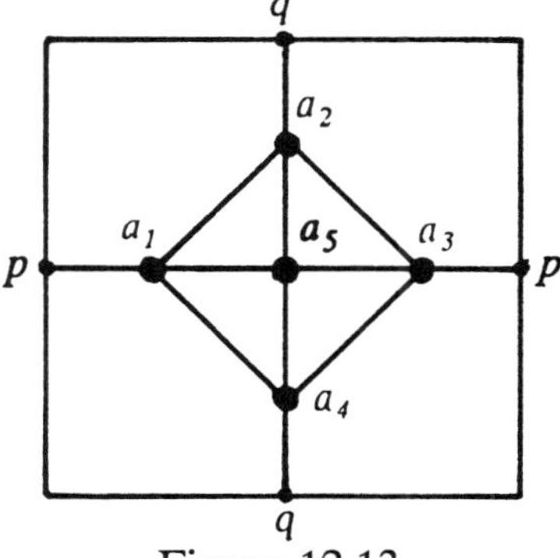

Figure 12.13

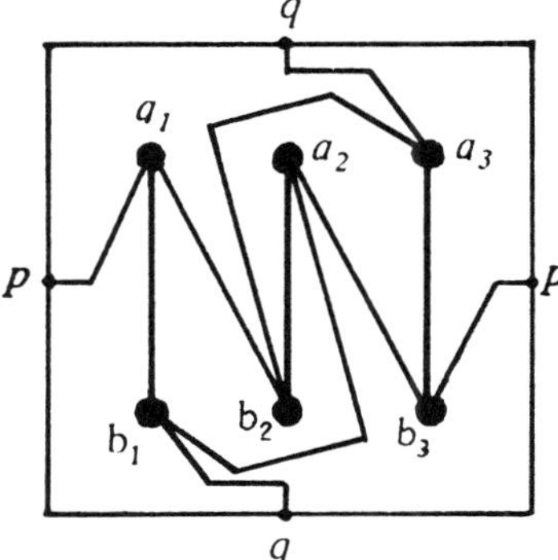

Figure 12.14

CHAPTER IV

IDEAS OF COMBINATORIAL GEOMETRY

13. WHAT ARE CONVEX FIGURES?

In this section we discuss (without proofs) some properties of *convex* figures in the plane.

A figure M is called **convex** if together with each two points a,b it contains the whole segment $[a,b]$ (Figure 13.1). Triangles, parallelograms, and trapezoids are convex (Figure 13.2). Disks, sectors with a central angle not exceeding 180°, and segments of

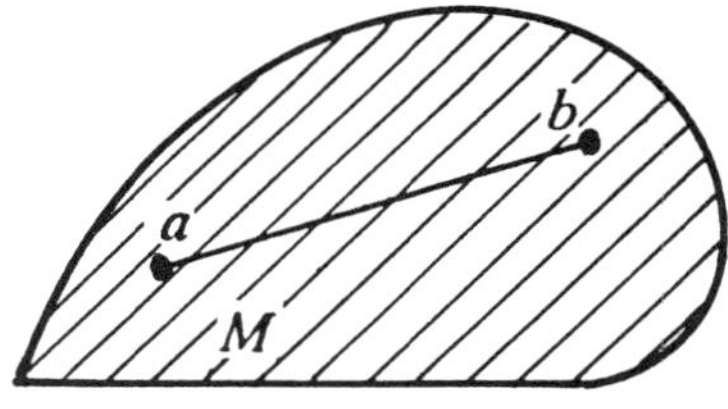

Figure 13.1

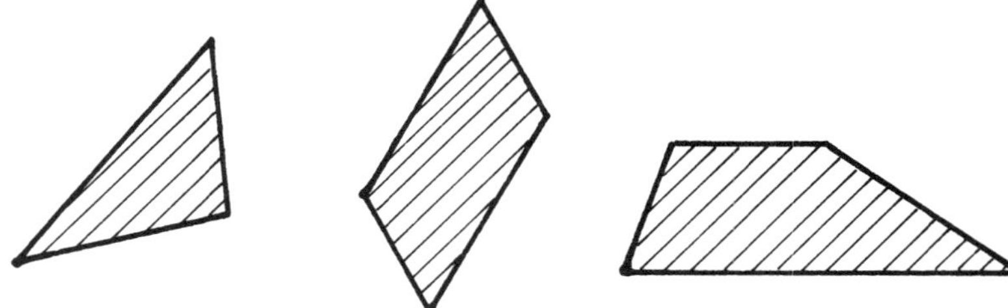

Figure 13.2

disks are also convex (Figure 13.3) and so is any segment, and any single point. In Figure 13.4 some non-convex figures are shown.

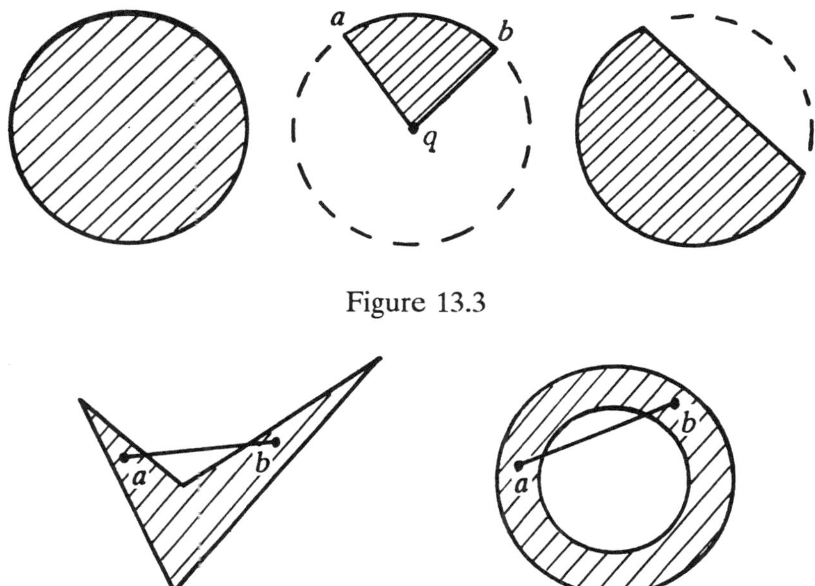

Figure 13.3

Figure 13.4

All figures mentioned above are bounded: each of them is contained in a disk (or in a square).

A *strip* is an example of an unbounded convex figure (Figure 13.5). As to angles, they may be either convex or non-convex. If an angle's measure does not exceed 180°, then the angle is convex (Figure 13.6); otherwise it is non-convex (Figure 13.7).

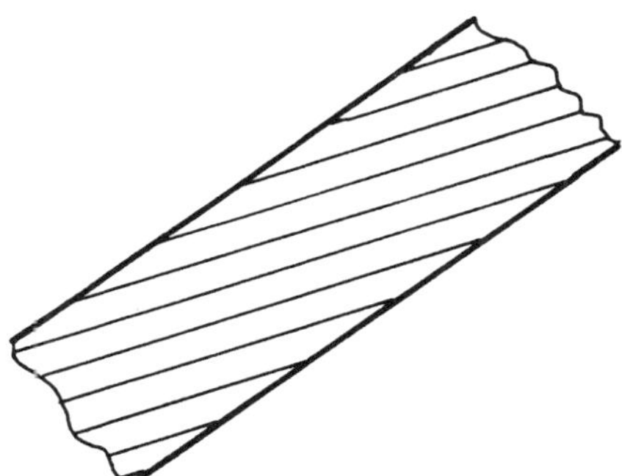

Figure 13.5

138

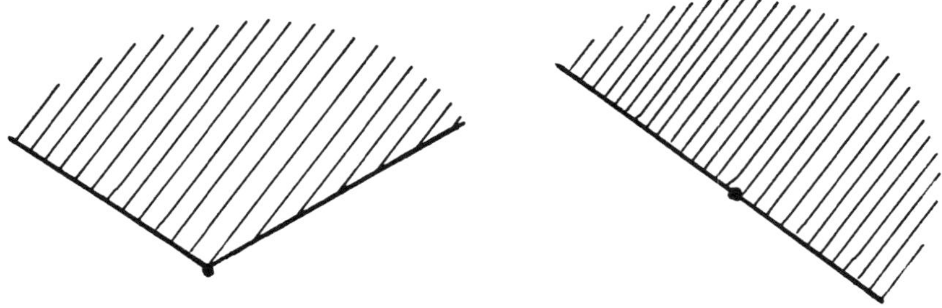

Figure 13.6

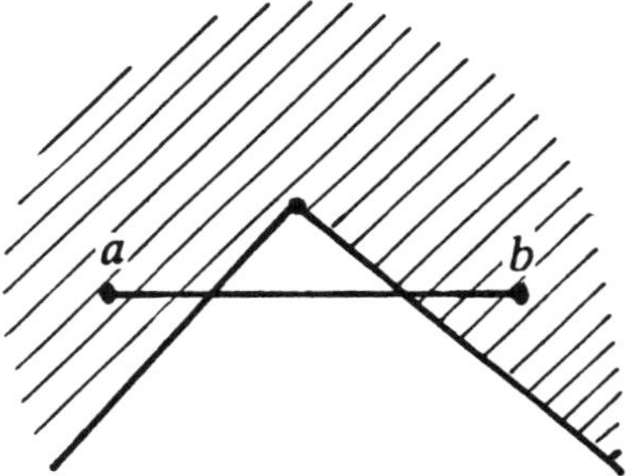

Figure 13.7

A point a of a triangle T is called a **boundary point** if any disk with the center a contains a point x that belongs to T, as well as a point y that does not belong to T (Figure 13.8). A point a of a triangle T is its boundary point if and only if a belongs to one of the sides of T. All other points of T are its **interior points**. For each interior point b of T (Figure 13.9), there exists a disk D with

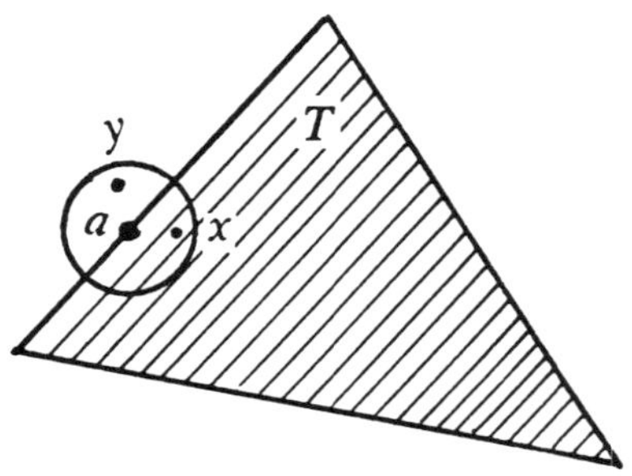

Figure 13.8

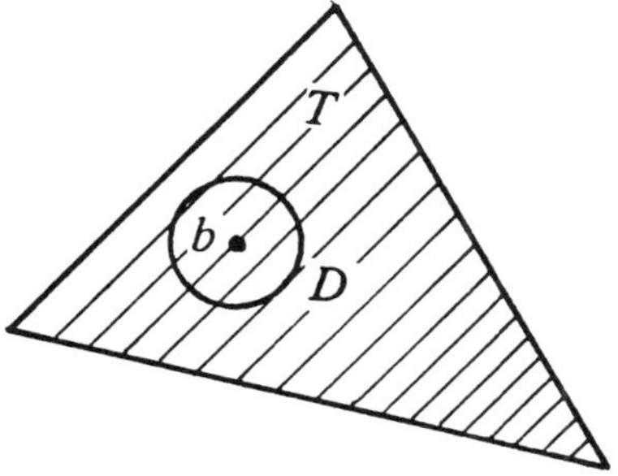

Figure 13.9

center b (maybe very small), such that D is completely contained in T. The above definitions are related not only to a triangle, but to any convex figure: any of its points can be either a boundary point (Figure 13.10) or interior point (Figure 13.11). A figure is

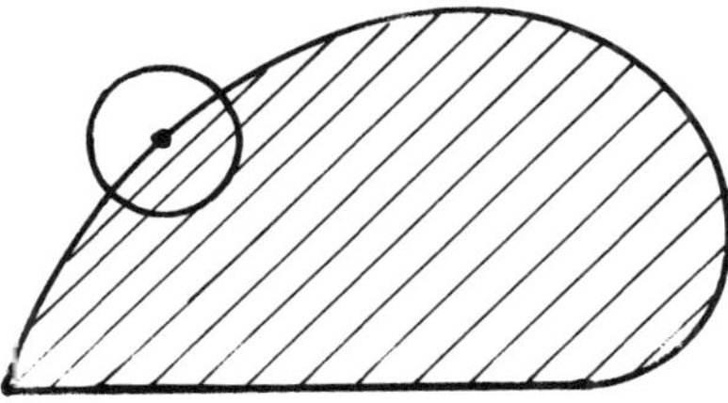

Figure 13.10

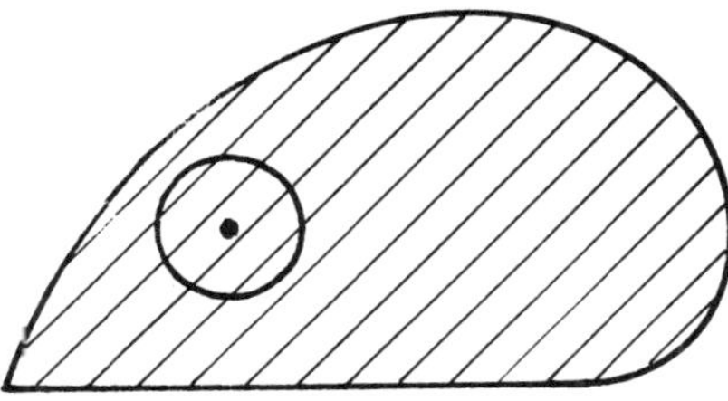

Figure 13.11

closed if it contains all its boundary points. **In the following sections 14-18 a "figure" would mean "a closed figure."** A convex figure that is closed and bounded is called *compact*. We will denote by *intM* the *interior* of a figure *M*, that is, the set of all interior points of *M* and by *bdM* its *boundary*, that is, the set of all boundary points of *M*.

Exercise 13.1. Let *a* be an interior point of a convex figure *M* and *p* a boundary point of *M*. Prove that all the points of the segment [a,p] except *p* are interior points of *M*.

Exercise 13.2. Prove that the intersection of two or more convex figures is a convex (possibly empty) figure.

Every line is a convex figure. Consequently, the intersection of a line *l* with a convex figure *M* is a convex figure situated on the line *l*, that is either a point, a segment, a ray, or the whole line (Figures 13.12 - 13.15).

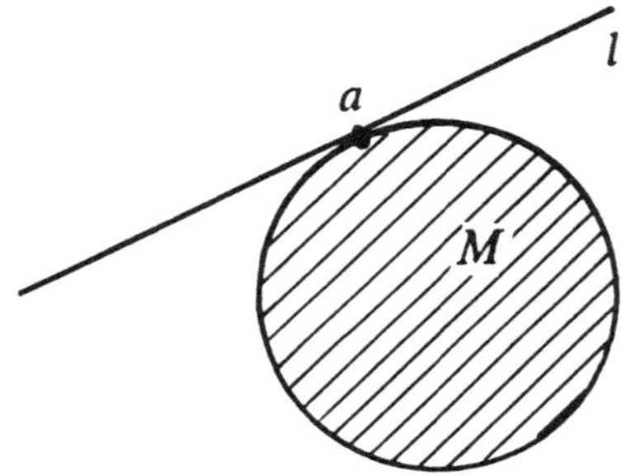

Figure 13.12

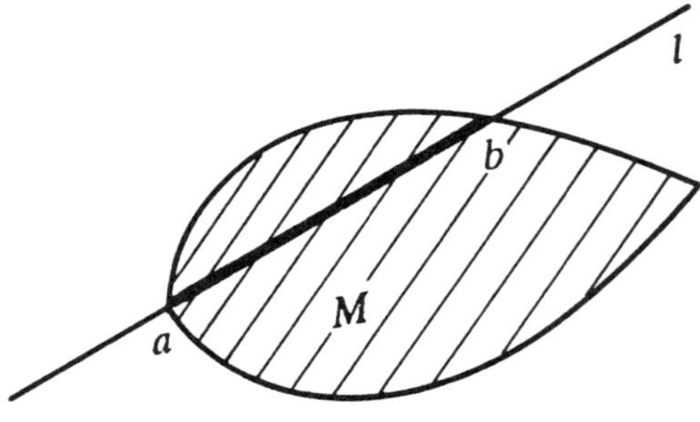

Figure 13.13

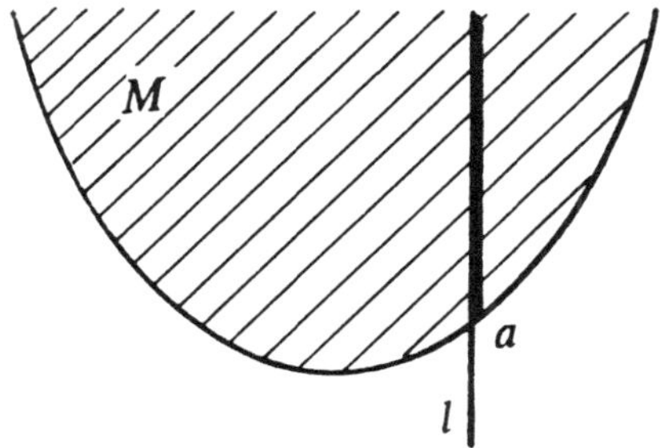

Figure 13.14

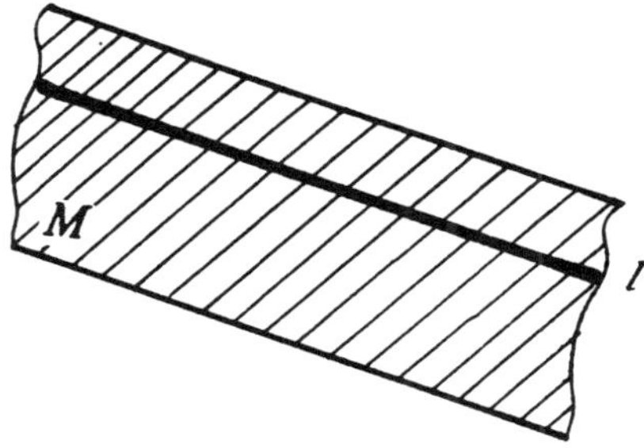

Figure 13.15

Exercise 13.2 implies that for every figure X (not necessarily convex) there exists the *smallest* convex figure that contains X. Indeed, if we consider *all* convex figures containing X, then their intersection is the smallest convex figure that contains X. This convex figure is called the **convex hull** of X and is denoted by *convX*.

A tight rubber band placed around a figure X gives a good visualization of this notion: it bounds the convex hull *convX* of X.

The convex hull of a finite point set $\{a_1 \ldots a_s\}$ in the plane (not all points on a line) is called a ***convex polygon***. If we have three points, not all on a line, then their convex hull is a triangle (Figure 13.16). As to the convex hull of four points, it can be either a triangle (Figures 13.17 and 13.18) or a quadrangle (Figure 13.19). In general, if we consider the convex hull of a finite set

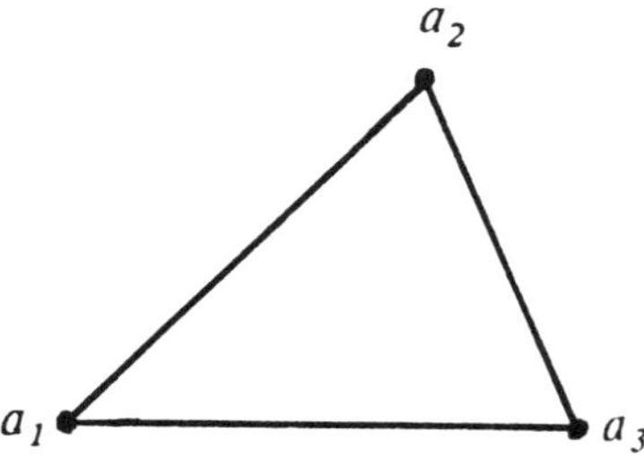

Figure 13.16

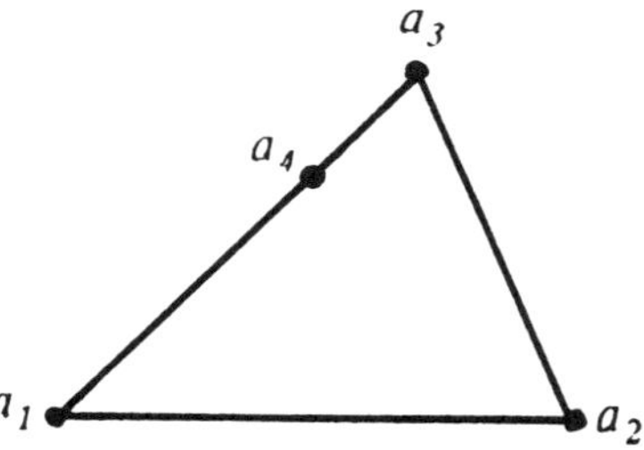

Figure 13.17

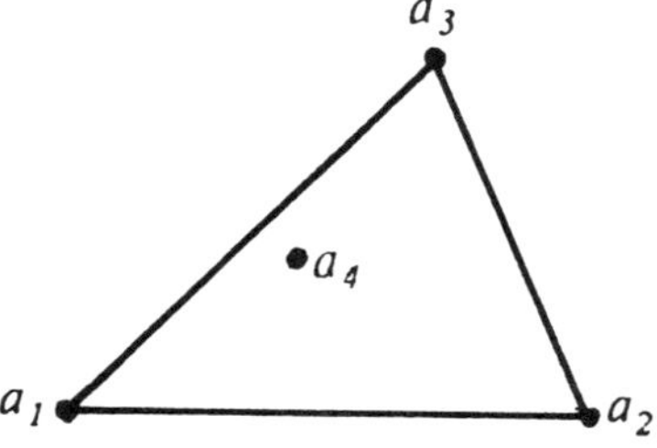

Figure 13.18

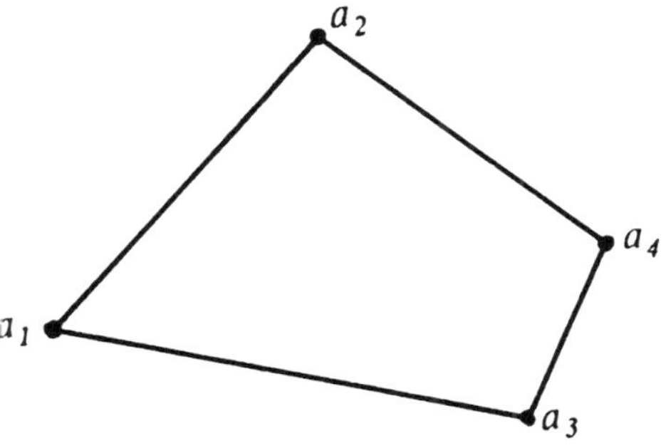

Figure 13.19

$\{a_1, ..., a_s\}$, then it is possible to permute subscripts and choose a minimal subset, say $\{a_1, ..., a_m\}$, $(m \leq s)$ such that $\mathrm{conv}\{a_1, ..., a_s\}$ $= \mathrm{conv}\{a_1, ..., a_m\}$. This means that the points $a_{m+1}, ..., a_s$ are situated on the boundary or in the interior of the convex polygon $M = \mathrm{conv}\{a_1, ..., a_m\}$ (Figure 13.20).

Similarly, the convex hull of a finite point set in the space (not all points in a plane) is called a *convex polyhedron.*

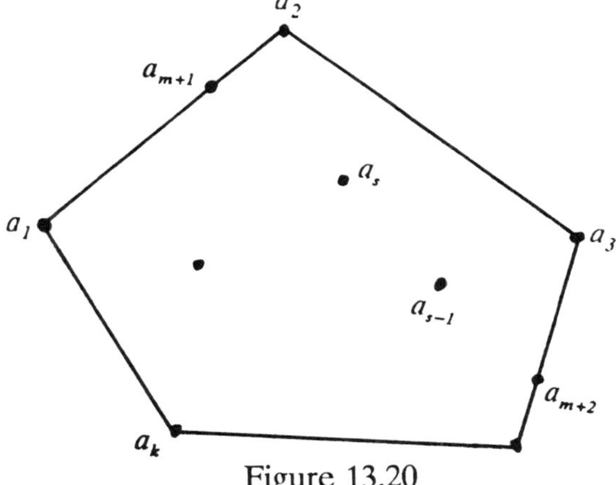

Figure 13.20

If $M = conv\{a_1...a_m\}$ is a convex polygon and the set $\{a_1, ..., a_m\}$ is minimal, then none of the points $a_1, ..., a_m$ belongs to the convex hull of all other points. These points are called *vertices* of the convex polygon M. Each vertex a_j of the convex polygon M is an *extremal point*, i.e., there exists no segment $[p,q]$ contained in M, such that a_j is an interior point of this segment (Figure 13.21). All other points of M are non-extremal (Figure 13.22). So, each

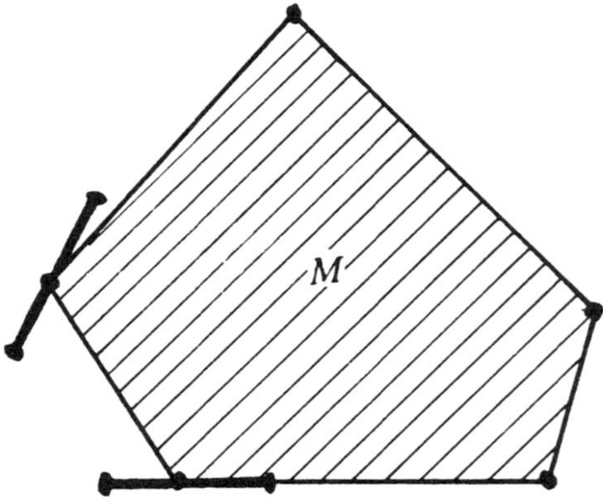

Figure 13.21

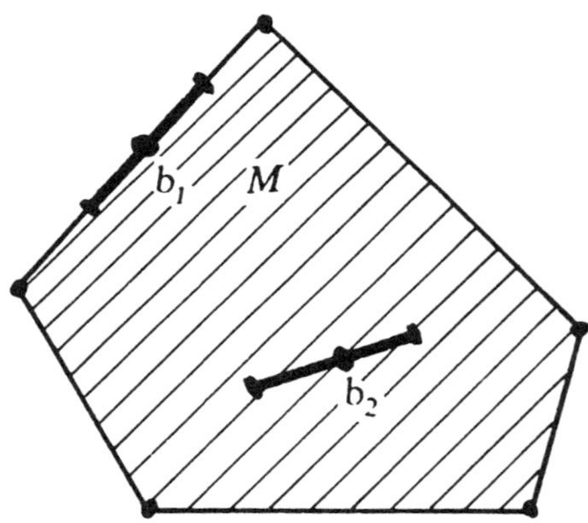

Figure 13.22

convex polygon is the convex hull of the set of its extremal points (and this set is minimal). This property of convex polygons has a generalization for all compact convex figures: each compact convex figure is the convex hull of the set of its extremal points. For a sector (Figure 13.3), the set of extremal points consists of its center q and all points of the arc ab.

Let F be a convex figure and L a line in the plane. We call L a **support line** of F if:

i) *L* and *F* have at least one common point and

ii) *F* is situated in a closed halfplane defined by *L* (Figure 13.23). If *F* is compact then the intersection of *F* and its support line *L* is either a point (Figure 13.23) or a segment (Figure 13.24).

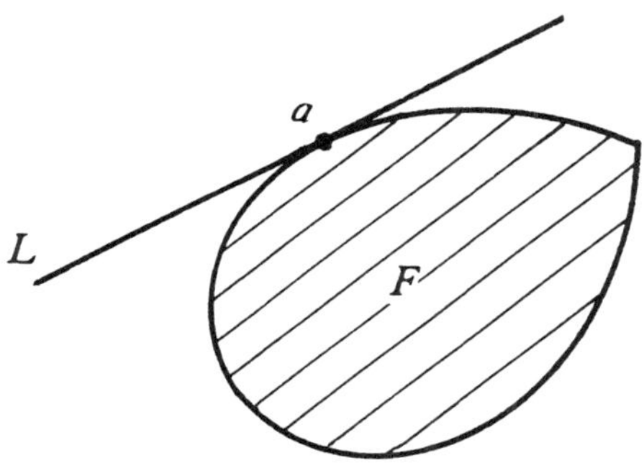

Figure 13.23

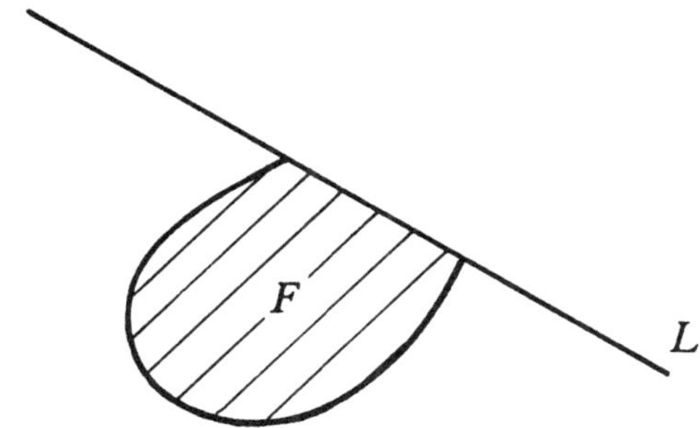

Figure 13.24

The following proposition plays an important role in geometry.

Theorem 13.1. *For every boundary point a of a convex figure F, there exists a support line of F passing through a.*

146

Conversely, if for every boundary point a of a closed plane figure F there exists a support line of F through a, then F is convex.

For example, if a is a vertex of a convex polygon F, then there are infinitely many support lines of M through a (Figure 13.25), and if a is a boundary point of M distinct from vertices, then there exists only one support line of M through a (Figure 13.26).

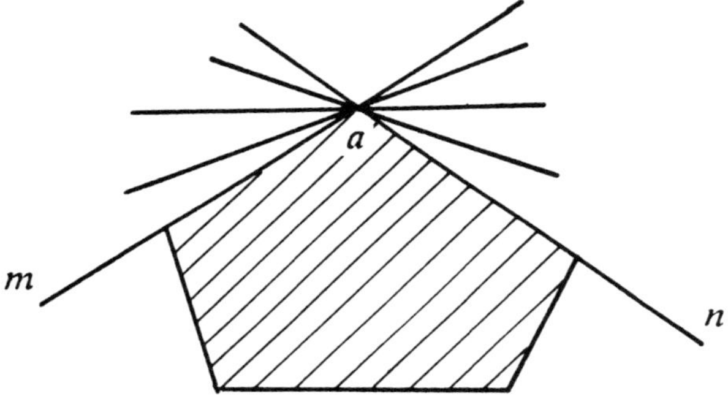

Figure 13.25

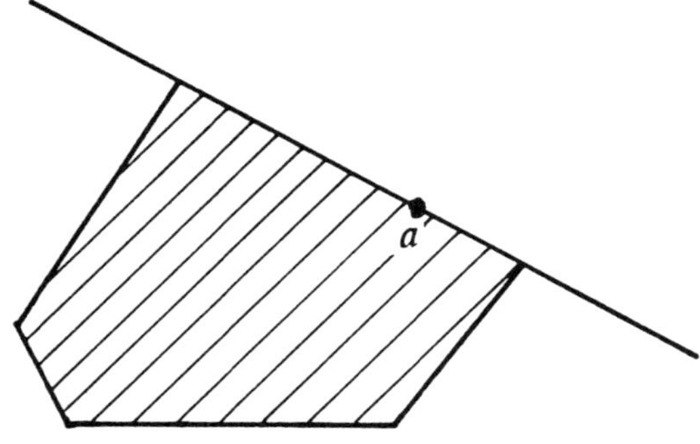

Figure 13.26

If there exists only one support line through a boundary point a of a convex figure F (Figures 13.26 and 13.27), then a is

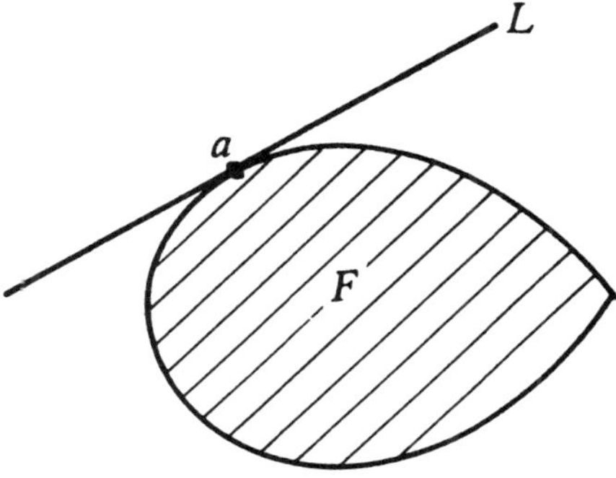

Figure 13.27

said to be a **regular boundary point** of F. Otherwise, (that is, when there are infinitely many support lines through a, (Figures 13.25 and 13.28), a is said to be an **angular boundary point** of F. For each angle boundary point a of a convex figure F there are two *tangent rays* of F emanating from a (the rays am and an in Figures 13.25 and 13.28). The Figure F is contained in the angle between these tangent rays.

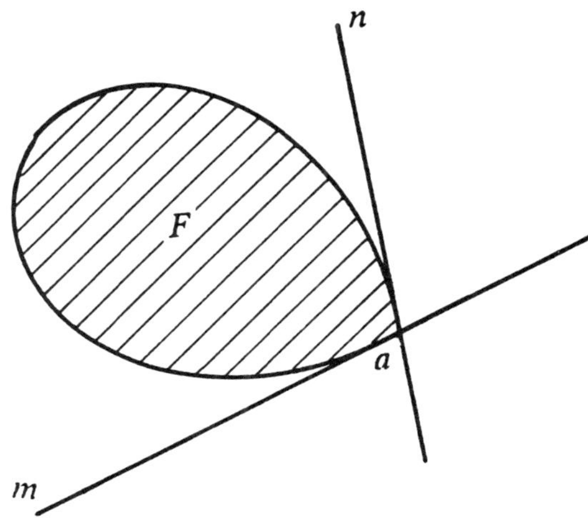

Figure 13.28

Exercise 13.3. Prove that the figure M bounded by parts of two congruent circles and their two common exterior tangent lines (Figure 13.29) is convex.

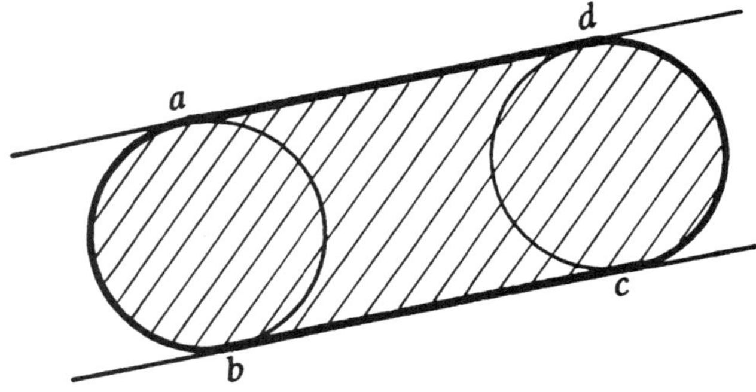

Figure 13.29

Exercise 13.4. Prove that if M is convex, then its ε-extension (Figure 13.30) is also convex (see the definition in Section 8, pages 77-78).

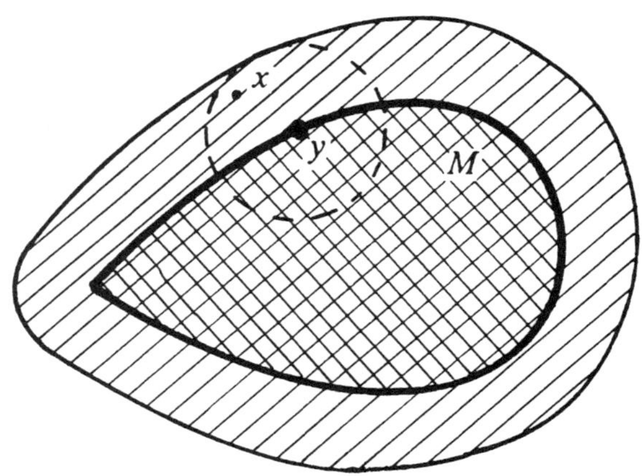

Figure 13.30

Exercise 13.5. Let M be a convex polygon of perimeter d and V the ε-extension of M. What is the length of the boundary of V?

Now we can state a very important theorem that is due to the German mathematician Wilhelm Blaschke:

Theorem 13.2. *(Blascke Theorem)* *Every infinite sequence of compact convex figures that is bounded, contains a subsequence that converges to a compact convex figure.*

Proof: The reader has probably noticed that this theorem is a direct consequence of Theorem 8.1. Indeed, we have to prove only that the limit M of a convergent sub-sequence F_1, F_2 ... of compact convex figures is compact and convex. Let us denote by d_i the distance between the compact figures M and F_i. Then the sequence d_1, d_2, ... of positive numbers converges to 0. For every $i = 1, 2, ...$ we denote by V_i the d_i-neighborhood of F_i. Then V_1, V_2, ... are compact convex figures (see Exercise 13.4). Moreover, M is the intersection of all figures V_1, V_2, ... Hence, M is convex and compact.

Exercise 13.6. Prove that for every compact convex figure F there exists a circumscribed circle, that is, the smallest circle that contains F. Is this circle unique?

Exercise 13.7. Prove that for every compact convex figure F there exists an inscribed circle, that is, the largest circle that is contained in F. Is this circle unique?

Finally, we consider *separability* of convex figures. Convex figures F_1 and F_2 in the plane are said to be **separable** if there exists a line L, such that F_1 and F_2 are situated in different closed halfplanes defined by L (Figure 13.31). Let us note that F_1 and F_2 may have a common boundary point (Figure 13.31) or even a common segment (Figure 13.32). If there is a line L, such that F_1 and F_2 are situated in the corresponding *open* halfplanes, then F_1 and F_2 are said to be **strictly separable** (Figure 13.33).

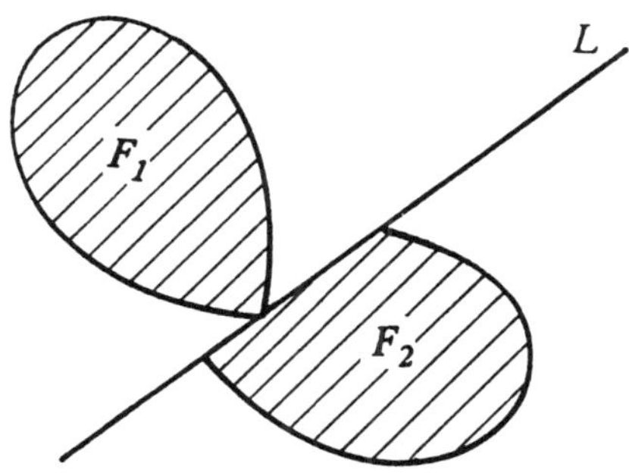

Figure 13.31

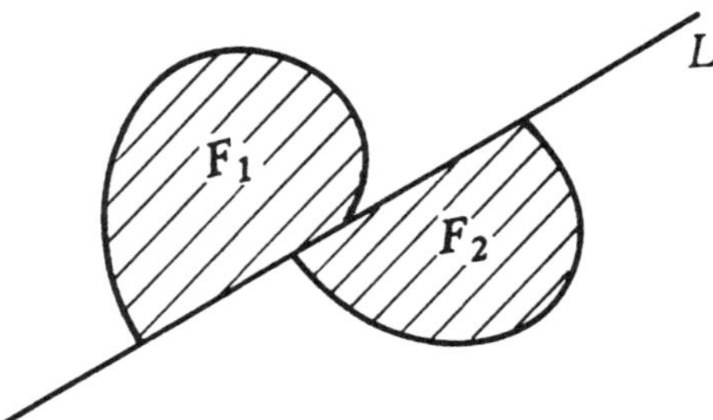

Figure 13.32

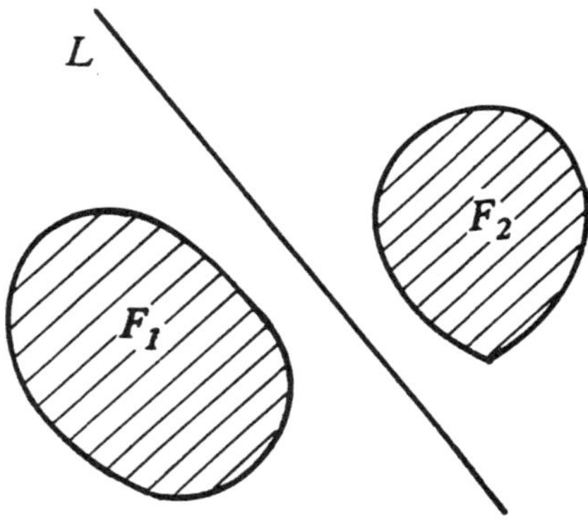

Figure 13.33

The following propositions indicate some sufficient conditions of separability for convex figures.

If F_1 and F_2 are figures in the plane with non-empty interiors that have no common interior points, then F_1 and F_2 are separable.

If F_1 and F_2 are closed convex figures in the plane without common points and at least one of them is compact, then F_1 and F_2 are strictly separable.

All the notions considered in this section can be generalized to n-dimensional case. For example, in place of support lines in the plane, we talk about **support planes** of a convex body in the space R^3.

Solutions of Exercises

13.1. Let b be a point on the segment $[a,p]$, $b \neq p$. Further, let A be a disk with center a contained in M and let r be its radius. Then there exists a disk B with center b contained in M. Figure 13.34 shows that we may take $\dfrac{|bp|}{|ap|}\, r$ as the radius* of B.

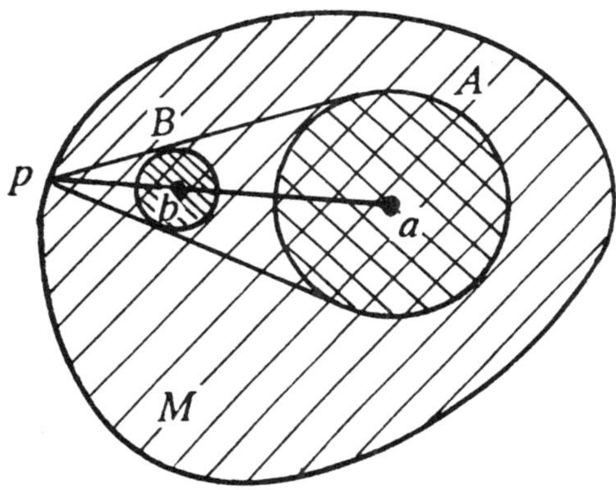

Figure 13.34

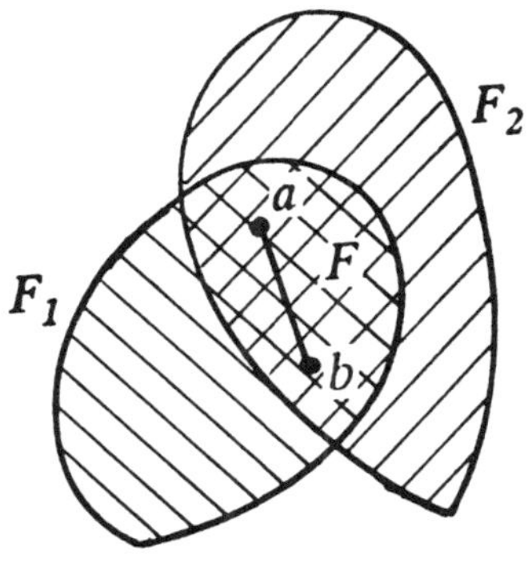

Figure 13.35

---

* The symbol $|bp|$ denotes the length of the segment $[b,p]$.

13.2. Let F_1 and F_2 be convex figures. We denote by F their intersection (Figure 13.35). Let a and b be points of F. Then a and b belong to the figure F_1 as well as to F_2. Since F_1 is convex, all points of the segment $[a,b]$ belong to F_1. The same is true for F_2. Consequently all points of $[a,b]$ belong to F. This means that F is convex.

13.3. The boundary of M consists of two semicircles ab and cd and two segments $[a,d]$ and $[b,c]$ (Figure 13.36). If x is a point of the arc ab (or of the arc cd), then the tangent line of this arc is a support line of M. Further, if y is a point of the segment $[a,d]$, then the line containing this segment is a support line of M (similarly for the segment $[b,c]$). Thus, for every boundary point of M there exists a support line of M through this point. This means, according to Theorem 13.1, that M is convex.

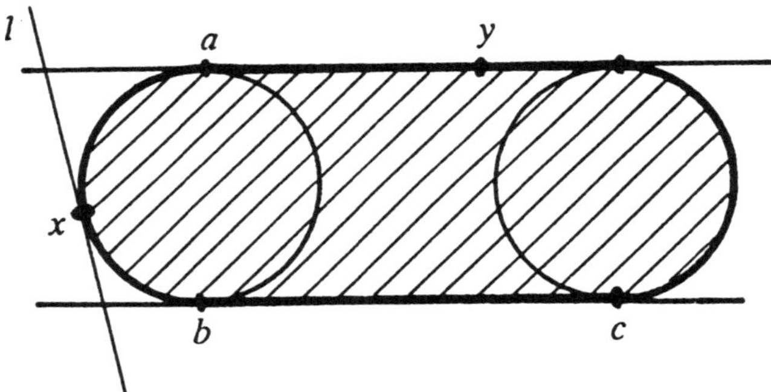

Figure 13.36

13.4. Denote by N the ε-extension of M (Figure 13.37). Let a and b be two points of N. Then there exist points x and y in M, such that the disk X of radius ε with center x contains a, and the disk Y of radius ε with center y contains b. Since M is convex, then the segment $[x,y]$ is contained in M. Consequently, any disk Z of radius ε with center z on the segment $[x,y]$ is contained in N. But the union of all such disks Z coincides with the convex hull of the figure consisting of two disks X and Y (see Exercise 13.3). Thus, this convex hull is contained in N. In particular, the segment $[a,b]$ is contained in N. This completes the proof.

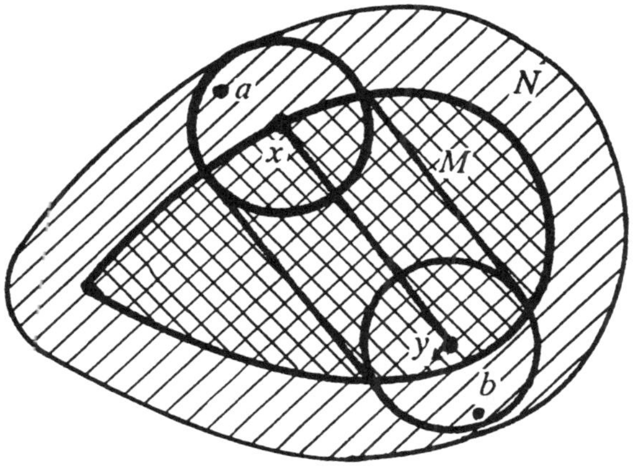

Figure 13.37

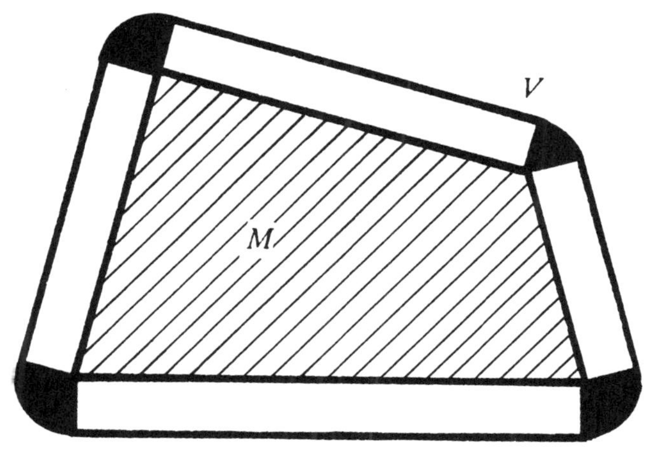

Figure 13.38

13.5. The boundary of V consists of segments, each of which is congruent to the corresponding side of M , and arcs of circles that taken together create a whole circle of radius ε (Figure 13.38. Consequently, the perimeter of V is equal to $p + 2\pi\varepsilon$.

154

13.6. We consider all the disks that contain F. Let r_0 be the exact lower bound of their radii. Then for every positive integer m there exists a disk D_m containing F with the radius less than $r_0 + \frac{1}{m}$. We obtain the sequence $D_1, D_2, ..., D_m, ...$ of disks. According to the Blaschke Theorem, a subsequence of this sequence converges to a compact convex figure D_0. Now it is clear that D_0 is a disk and its radius is equal to r_0. Hence, D_0 is the minimal radius disk containing F, that is, D_0 is the disk circumscribed about F.

If there were another disk D'_0 of radius r_0 containing F, then F would be contained in the intersection of the disks D_0 and D'_0. But this intersection as well as the figure F is contained in a disk of radius *smaller* than r_0 (bold circle in Figure 13.39), contradicting the definition of the number r_0. Thus, the circumscribed disk is unique.

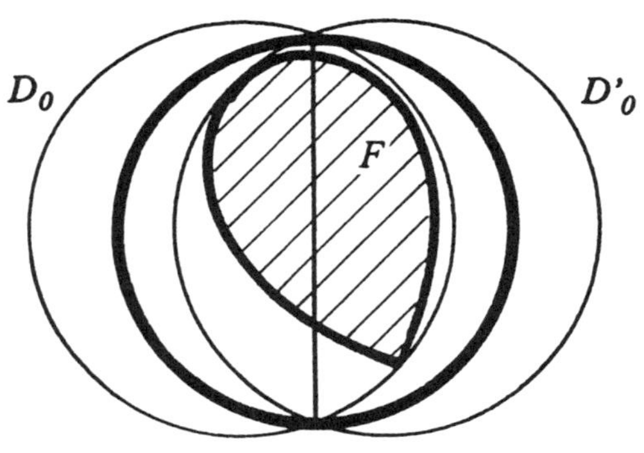

Figure 13.39

13.7. The solution is similar to the solution of Exercise 13.6, but in this case uniqueness does not hold! This is shown in Figure 13.40. However, *if the boundary of F contains no segment, then the inscribed circle is unique.* We do not wish to deny the readers the pleasure of discovering the proof of it on their own.

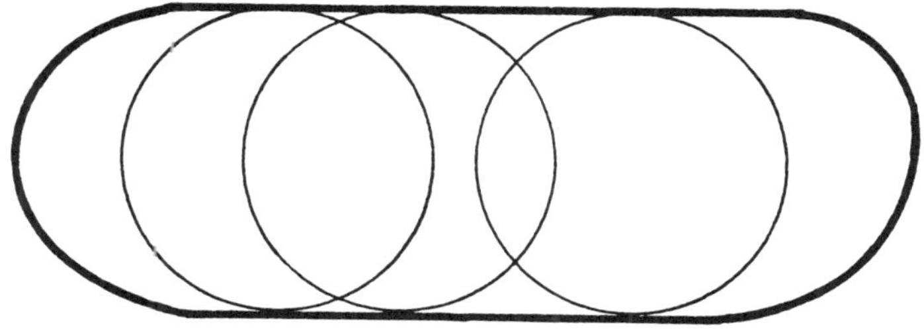

Figure 13.40

14. DECOMPOSITION OF FIGURES INTO PARTS OF SMALLER DIAMETERS

The *diameter* of a figure is the greatest distance between its points. More precisely, a figure F has diameter d if

i) any two points of F are at a distance at most d; and

ii) there are two points x and y in F whose distance is equal to d.

The problem of decomposition into parts of smaller diameters was formulated by the Polish mathematician Karol Borsuk in 1933. The birth of this problem was preceded by three mathematical events. The first of them is described in the following two exercises.

Exercise 14.1. Prove that a disk of diameter d can be decomposed into three parts of smaller diameters, but cannot be decomposed into two parts of smaller diameters.

Exercise 14.2. Prove that a ball of diameter d can be decomposed into four parts of smaller diameters.

We denote the plane by R^2 and the three-dimensional space by R^3. Then we can combine the results of Exercises 1 and 2 in the following statement:

The ball in n-dimensional Euclidean space R^n (where $n = 2$ or 3) can be decomposed into $n + 1$ parts of smaller diameters.

It is not so difficult to prove (if you are familiar with the notion of n-dimensional space for arbitrary natural number n) that this fact holds for *any* finite-dimensional Euclidean space. But the result of Exercise 14.1 contains more. It affirms that for $n = 2$ the n-dimensional ball cannot be decomposed into n parts of smaller diameters. Is it true for arbitrary n? This question brought about the second mathematical event: In 1932 Karol Borsuk proved that for any $n \geq 2$ an n-dimensional ball of diameter d cannot be decomposed into n parts of smaller diameters. Note that the same result was obtained by two Russian mathematicians, Lazar Lusternik and Lev Schnirelmann in 1930. K. Borsuk did not know about that and obtained the result independently (and with a different proof).

The third event occurred in 1933. K. Borsuk proved the following proposition:

Theorem *14.1*. (Borsuk Theorem) *Every figure of diameter d in R^2 (not only a disk) can be decomposed into three parts of smaller diameters.*

We are going to sketch a proof of this result, following Borsuk's ideas. He used an interesting geometrical result obtained in 1920 by an Hungarian mathematician, J. F. Pál. Here is Pál's result:

Any plane figure F of diameter d can be inscribed into a regular hexagon with distance d between its opposite sides (Figure 14.1).

Here is the idea of the proof: We draw an initial ray m_0 from which we are going to measure angles. Now we construct two parallel support lines L_1 and L_2 of the figure F that make the angle ϕ_0 with the initial ray m_0 (Figure 14.2). The distance between L_1 and L_2 does not exceed d (can you see why?). If this distance is less than d, then we move these lines apart until the distance between them is equal to d (dotted lines in Figure 14.2).

Thus, for every angle ϕ there is a strip of width d containing F. We denote this strip by $S(\phi)$.

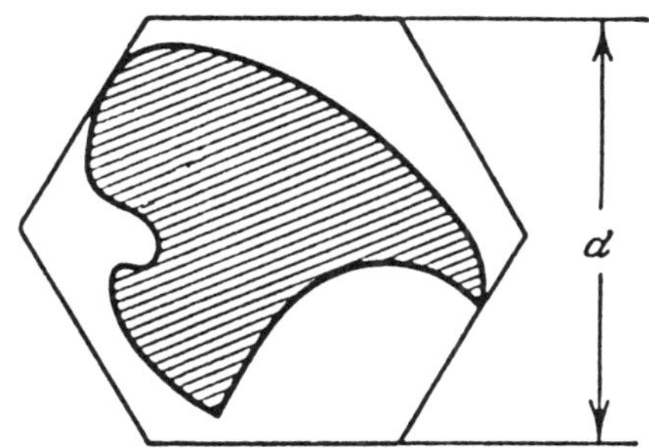

Figure 14.1

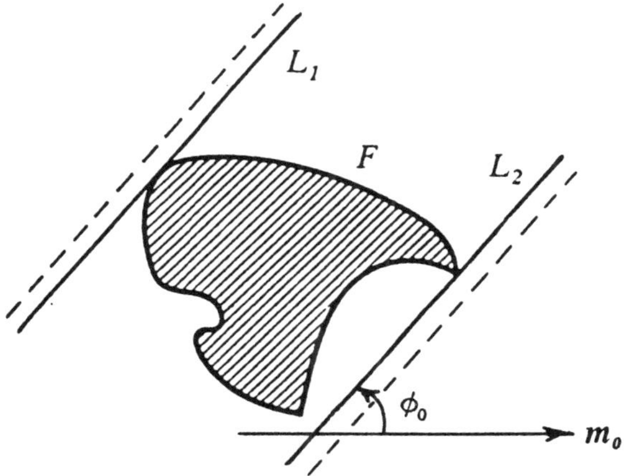

Figure 14.2

Now we construct the strips $S(\phi_0)$ and $S(\phi_0 + 60°)$. Their intersection is the rhombus containing F that has an angle of 60°, and distance d between its opposite sides (Figure 14.3). Finally, we construct the strip $S(\phi_0 + 120°)$ (dotted lines in Figure 14.3). The intersection of three strips is the hexagon $H(\phi)$ with angles of 120° containing F. This hexagon is regular if the lengths of altitudes h_1 and h_2 drawn to the dotted lines in Figure 14.3 are equal: $h_1 = h_2$, and irregular otherwise. Let us assume that $h_1 \neq h_2$ (say, $h_1 < h_2$). Now we will increase the angle ϕ_0 to $\phi_0 + 180°$. Then h_1 and h_2 will change continuously with ϕ (this visually clear assertion requires a rigorous proof), and consequently the difference $h_1 - h_2$ will change continuously. But when ϕ reaches $\phi_0 + 180°$ the lengths h_1 and h_2 will switch their roles. So, the difference, being negative at first, becomes positive in the end. Hence, by Intermediate Value Theorem, there exists an angle $\phi = \phi$, such that $h_1 - h_2 = 0$. At this angle ϕ_1 the hexagon $H(\phi_1)$ is regular.

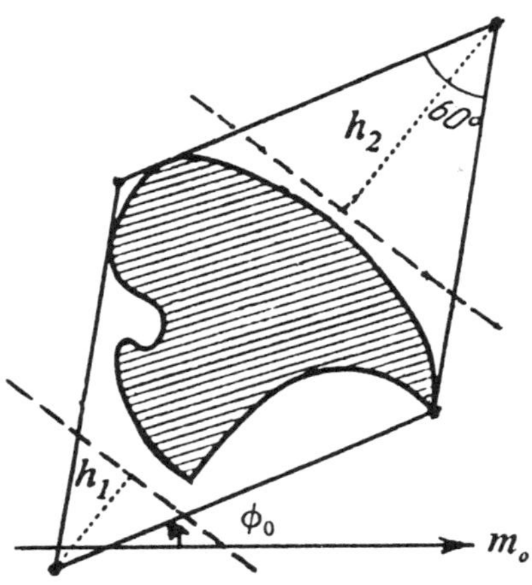

Figure 14.3

Exercise 14.3. Prove Theorem 14.1 with the help of the above result of Pál.

Let us now introduce a useful notation. For a plane figure F of diameter d we denote by $a(F)$ the smallest of the integers s, such that there exists a decomposition of F into s parts of smaller diameters. We will call $a(F)$ the **Borsuk number** of F.

The three events mentioned above can be now combined into the following two assertions:

i) $a(B) = n + 1$ for an n-dimensional ball B

ii) $a(F) \leq 3$ for every bounded plane figure F

These assertions led K. Borsuk to formulate the following hypothesis:

The Borsuk Conjecture. *For any bounded figure F in R^n, the inequality $a(F) \leq n + 1$ is true.*

What is the state of this problem today? In 1955, the English mathematician H. G. Eggleston proved that in R^3 the Borsuk conjecture is true.

Theorem 14.2 (H.G. Eggleston [E1]) *Every body of diameter d in R^3 can be decomposed into four parts of smaller diameters.*

Eggleston's proof was rather complicated. In 1957, the Israeli (now American) mathematician Branko Grünbaum [G1] and the Hungarian mathematician A. Heppes [Hep] found other proofs that were much simpler. The proofs had similar main ideas: to find a "more or less small" universal cover W for all three-dimensional bodies of diameter d (which means that every body of diameter d can be embedded into W) and decompose this cover into four parts of diameters not exceeding d. So, this idea is a direct generalization of the method that was applied in the proof of Theorem 14.1 (see Exercise 14.3). Grünbaum's paper was written in English and Heppes' in Hungarian. This is why the first paper is better known. Grünbaum also gives a better estimate: his four parts of the universal cover have diameters $\approx 0.9887d$ whereas Heppes' estimate is $\approx 0.9977d$.

Let us sketch the Grünbaum construction. As David Gale proved in [Ga], *every body of diameter d can be embedded in a regular octahedron with distance d between its opposite faces.* This fact can be proved similarly to Pál's theorem. So, this octahedron is a universal cover for all bodies of diameter d. Further, Grünbaum has noticed that it is possible to cut off three vertices of this octahedron by planes that are situated at distance $\frac{d}{2}$ from the center of the octahedron (see Figures 14.4, 14.5, and 14.6), and the truncated body remains a universal cover. Finally, Grünbaum decomposes this truncated body into four parts (Figure 14.7) and computes the diameters of these parts.

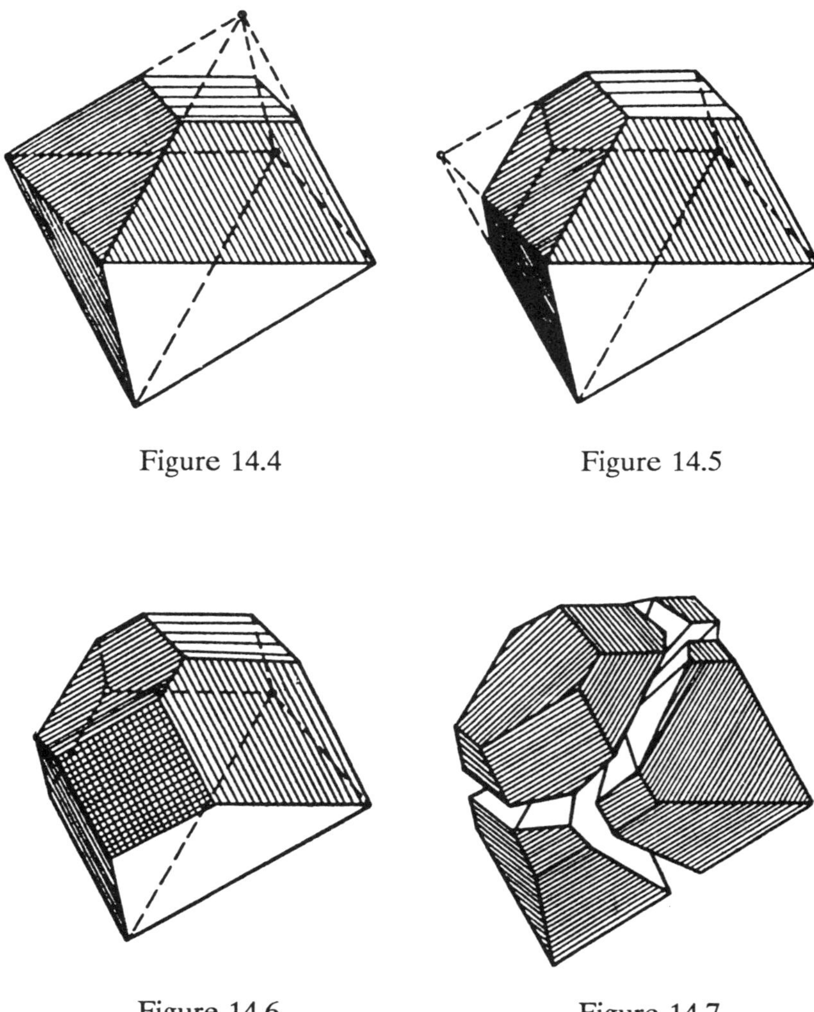

Figure 14.4   Figure 14.5

Figure 14.6   Figure 14.7

And he was lucky! It turns out that if the decomposition is made carefully, then all the diameters are equal to

$$\frac{\sqrt{6129030 - 937419\,\sqrt{3}}\;d}{1518\sqrt{2}} \approx 0.9887\,d$$

The reader can find a detailed account of this proof in the book by Vladimir Boltyanski and Israel Gohberg [BG]. Many times, from different countries and cities mistaken solutions of the

Borsuk problem were sent to the authors of the book [BG]. One such "proof" was even found by a member of the Soviet Academy of Sciences. This shows how difficult the Borsuk problem is in general.

It is still open for $n \geq 4$. Is there in R^4 a convex polytope M (i.e., a convex hull of a finite point set) such that $a(M) = 6$? Perhaps such a polytope can be found with the help of computers? At any rate, the authors are inclined to believe that a counter example exists to the Borsuk problem in the case $n \geq 4$.

And now we will mention some partial results in the direction of Borsuk's conjecture and its generalizations.

First of all let us return to the plane R^2. Not for every plane figure F do we have $a(F) = 3$. For example, for a parallelogram F, we have $a(F) = 2$ (Figure 14.8). Note that $a(F) \geq 2$ for any figure F. Thus, a new problem naturally arises: for what figures F in R^2 the equality $a(F) = 3$ is true? And for what figures F, $a(F) = 2$? This problem was solved by V. G. Boltyanski in 1970 ([B2]). We will discuss this solution in Section 16.

Further, for convex bodies F in R^n with a smooth boundary the Borsuk problem has a positive solution. This result will be discussed in Section 17.

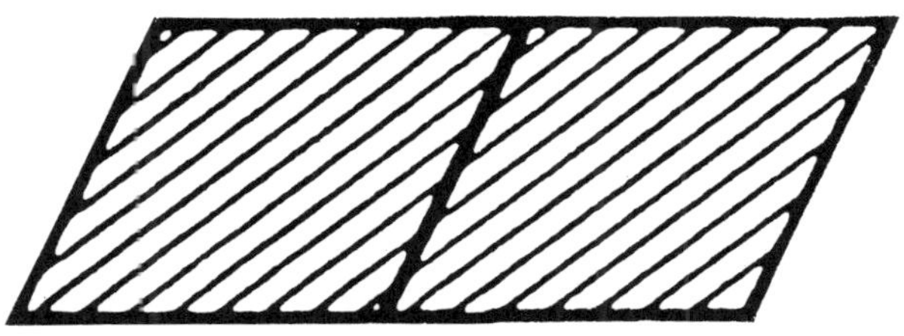

Figure 14.8

Solutions of Exercises

14.1. A decomposition of a disk into three parts of smaller diameters is shown in Figure 14.9. Another solution is given in Figure 14.10.

Figure 14.9

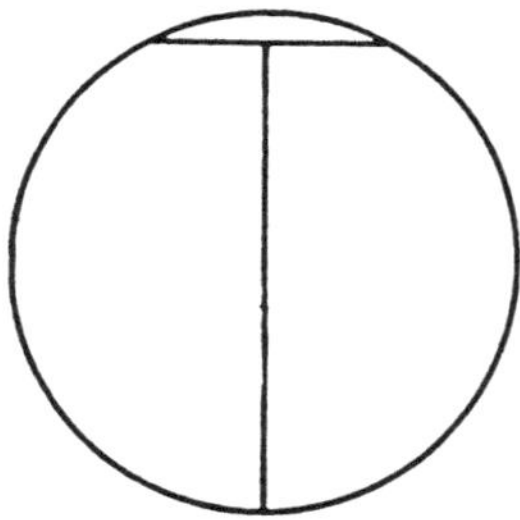

Figure 14.10

Now let us assume that there exists a decomposition of the disk F of diameter d into two parts A and B of smaller diameters. We can assume without loss of generality that both sets A and B are closed because the diameter of a set is equal to the diameter of its closure. If *the boundary bdF* of F is contained in A, then the diameter of the part A is equal to d, contradicting the choice of the decomposition. So, there exists a point b in *bdF* that does not belong to A. That is, b belongs to B. Similarly, there exists a point a in *bdF* that belongs to A. Since *bdF* is connected and the sets A,B are closed, there exists a point c in *bdF* such that c belongs to both A and B. Let now c' be a point in *bdF* that is diametrically opposite to c. If c' is in A, then the diameter of A is equal to d because c and c' both belong to A (Figure 14.11). And if c' is in B, then the diameter of B is equal to d because c

and c' both belong to B (Figure 14.12). This contradiction shows that there exists no decomposition of the disk F into two parts of smaller diameters.

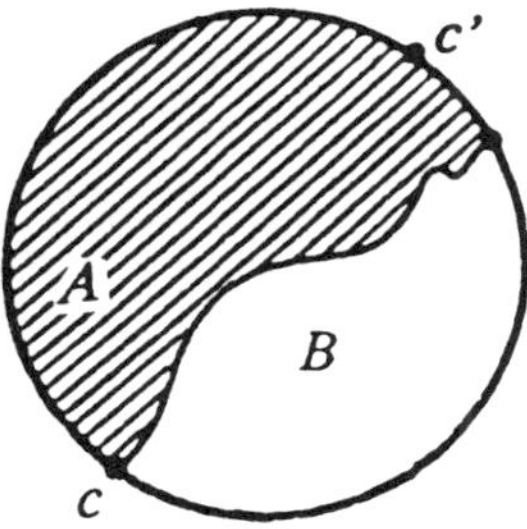

Figure 14.11

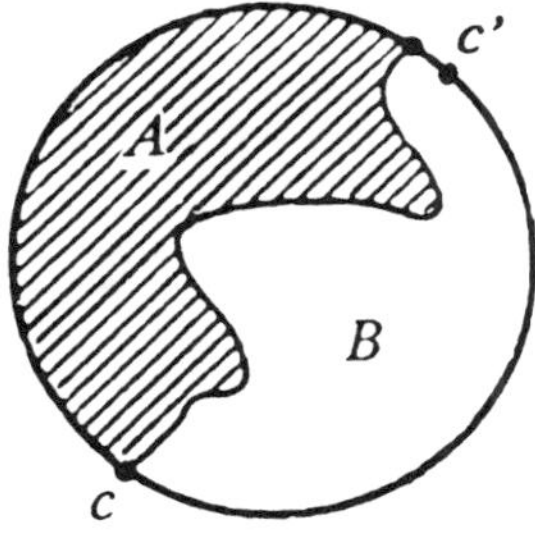

Figure 14.12

14.2. A solution is presented in Figure 14.13a. The upper cap is less than a hemisphere, and three planes cutting the lower part of the ball form three dihedral angles and meet each other at an angle 120°. The other solution in Figure 14.13b is more symmetric. We inscribe a regular tetrahedron into the ball and project its four faces onto the boundary sphere of the ball from its center. Each part of the ball is the convex hull of the center of the ball and one of the aforesaid projections.

Figure 14.13a   Figure 14.13b

14.3. A solution is shown in Figure 14.14. Indeed, the hexagon is decomposed into three parts, each of which has a diameter smaller than d. Hence, the figure F inscribed in the hexagon is decomposed too.

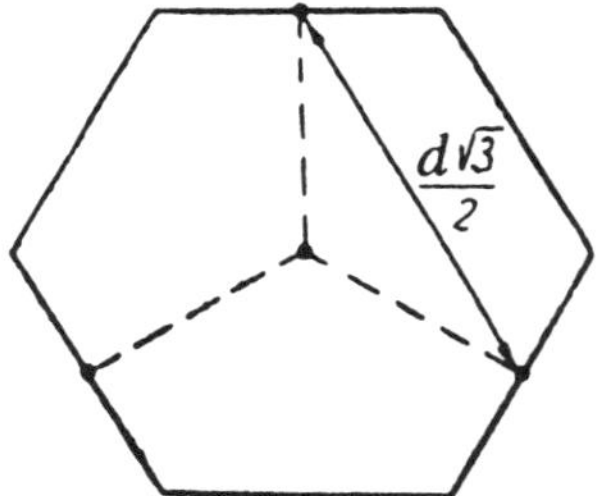

Figure 14.14

15. FIGURES OF CONSTANT WIDTH

Let F be a plane convex figure and L be a line in the plane. The distance h between two support lines of F parallel to L (Figure 15.1) is said to be the width of the figure F in the direction L. If the width of the figure F is the same in all the directions, then F is said to be **a *figure of constant width*.**

Certainly, every disk is a figure of constant width. But there is an infinite family of constant width figures different from disks (Figure 15.2).

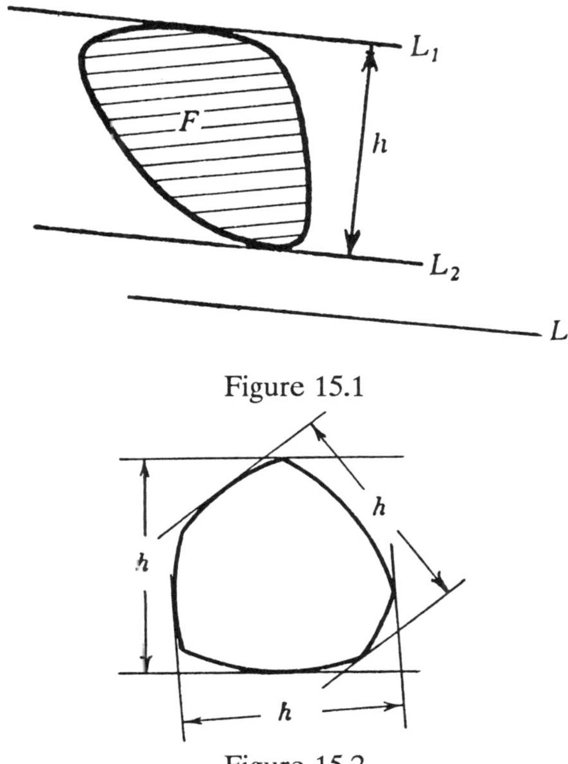

Figure 15.1

Figure 15.2

Exercise 15.1. *The Reuleaux Triangle* (Figure 15.3) is bounded by three arcs of circles with centers at the vertices of a regular triangle. Prove that this figure has a constant width.

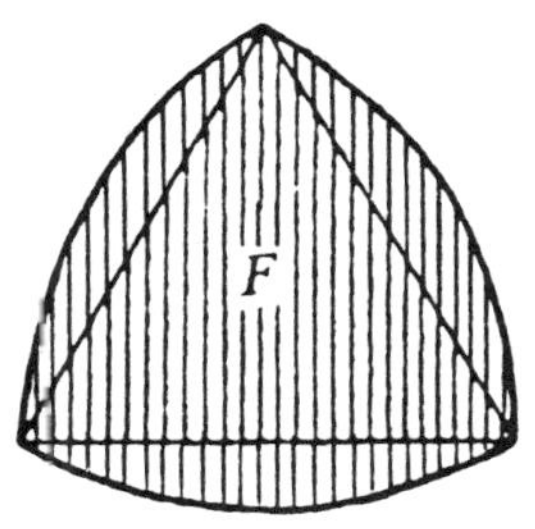

Figure 15.3

Exercise 15.2. Let M be a regular polygon with an odd number of vertices. We construct arcs with centers at the vertices of M joining two opposite vertices (Figure 15.4).

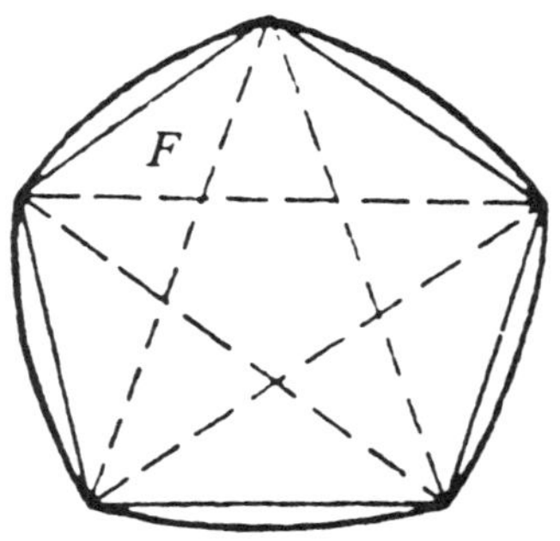

Figure 15.4

Prove that these arcs bound a figure of constant width. Generalize this result to the case when the polygon M is not regular (Figure 15.5), but each of its vertices is an endpoint of two diagonals of the same length h (and the lengths of other diagonals are less than h).

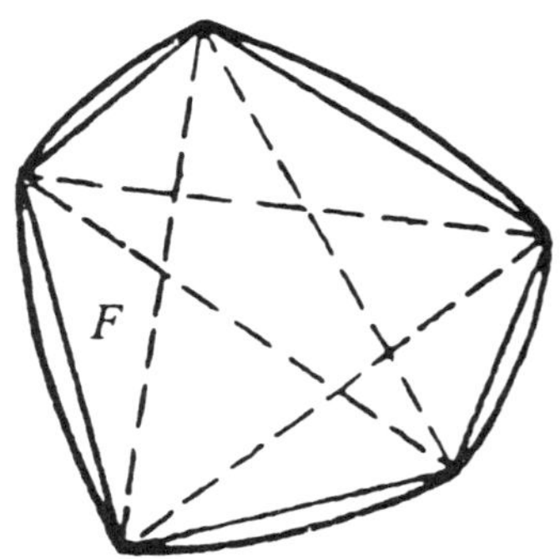

Figure 15.5

Exercise 15.3. Prove that the diameter d of a figure of constant width h is equal to h.

Let F be a figure of constant width h. Every chord of F that has length h is called a **_diametral chord_** of F (Figure 15.6).

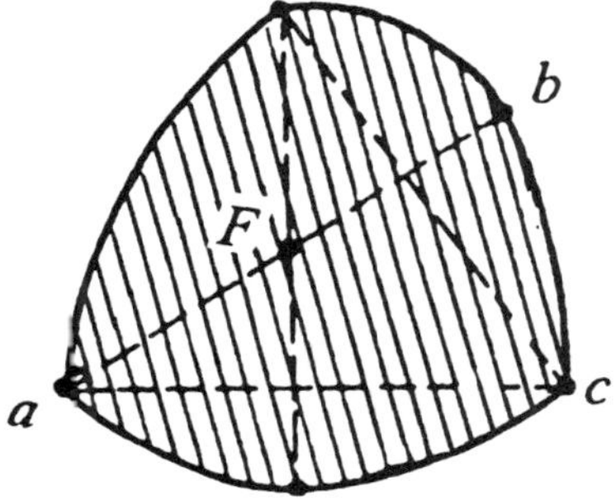

Figure 15.6

Exercise 15.4. Prove that each boundary point a of a figure F of constant width is an endpoint of at least one diametral chord of F.

Exercise 15.5. Let $[a,b]$ be a diametral chord of a constant width figure F and L,M be the lines that are perpendicular to $[a,b]$ and pass through a,b respectively (Figure 15.7). Prove that L and M are support lines of F.

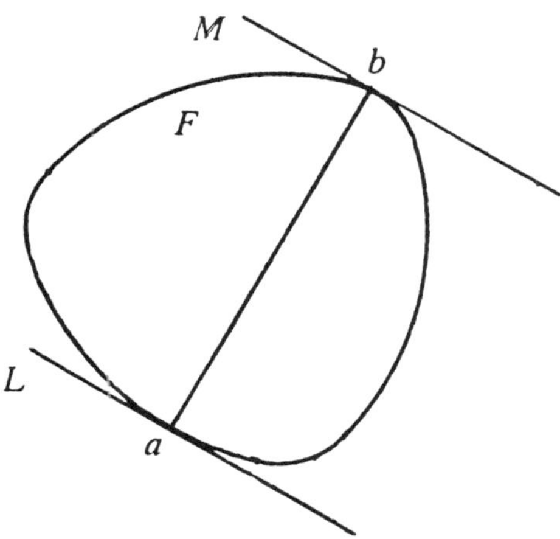

Figure 15.7

168

Exercise 15.6. Prove that every support line L of a constant width figure F has only one common point with F.

Exercise 15.7. Prove that every two diametral chords of a constant width figure F have a common point.

Exercise 15.8. Let F be a figure of constant width h and Q be a figure of diameter h that contains F. Prove that Q coincides with F.

The exercises above allow the reader to be on a friendly footing with constant width figures. Now we are going to consider some important and interesting properties of such figures. The first of them was established by the French mathematician E. Barbier.

Theorem 15.1. *(Barbier Theorem) The perimeter of every figure of constant width h is equal to πh.*

Proof. First of all we will formulate the following auxiliary proposition. Prove it on your own. Let $abcd$ be a rhombus and L,M be two parallel lines at the distance h from each other that intersect the rhombus and are perpendicular to the diagonal $[b,d]$ (Figure 15.8). Then the intersection of the rhombus and the strip between the lines L,M is a hexagon, whose perimeter does not depend upon the choice of the lines L,M.

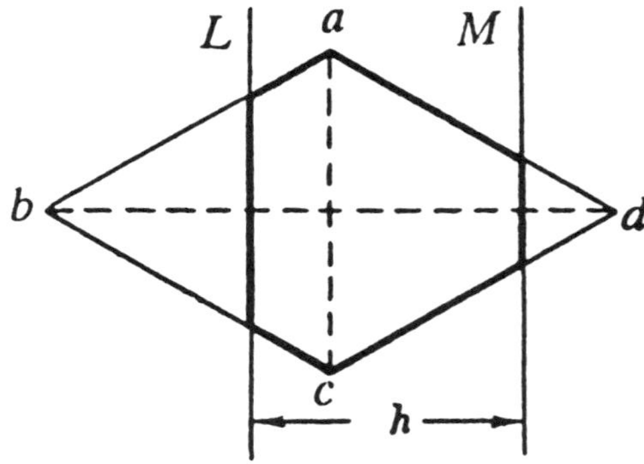

Figure 15.8

Now let F be a figure of constant width h and K be a disk of the same width h. It is clear that the squares circumscribed about F and K are congruent and, consequently, have the same perimeter. We can use this fact as the basis of induction.

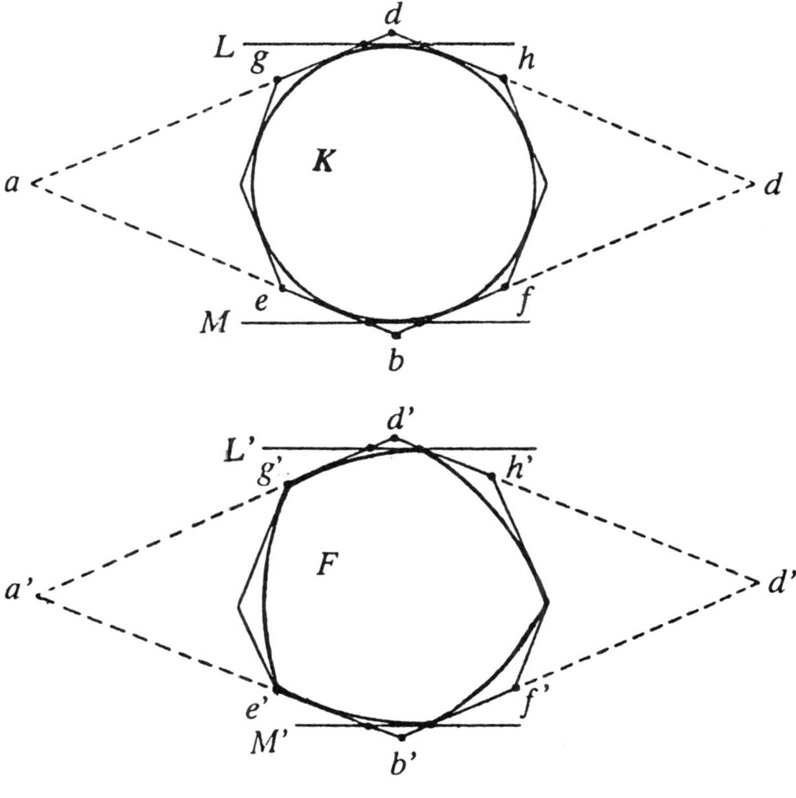

Figure 15.9

Assume that the 2^n-gons with equal angles circumscribed about F and about K have the same perimeter. Let us prove now that the same is true for 2^{n+1}-gons. Indeed, consider two adjacent sides $[b,e]$, $[b,f]$, and the opposite sides $[d,g]$, $[d,h]$ (Figure 15.9). The straight lines containing these sides form a parallelogram with both altitudes equal to h, i.e., rhombus. Similarly, we construct the rhombus for F. These rhombuses for K and F are congruent to each other (Figure 15.9). Now we construct support lines L,M for K that are perpendicular to the diagonal $[b,d]$ and similar support lines L',M' for F. The distance between L and M (and between L' and M' as well) is equal to h. According to the above auxiliary proposition, the obtained polygons (that is, the intersections of 2^n-gons with the constructed strips) have equal perimeters. Carring out this construction for each pair of adjacent sides, we obtain 2^{n+1}-gons with equal angles circumscribed about K and F and prove that these 2^{n+1}-gons have equal perimeters.

We showed that 2^n-gons with equal angles circumscribed about K and F have equal perimeters for every n. By increasing n without bound we can show that the boundaries of K and F have the same length. But for K this length is equal to πh. Consequently, the length of the boundary of F is equal to πh too.

In order to state the next theorem we have to introduce the operation of the **vector addition of convex figures**. Let F_1 and F_2 be convex figures in the plane, and assume the origin point is chosen. Each point a of the plane is identified with the vector starting at the origin and ending at the point a (Figure 15.10).

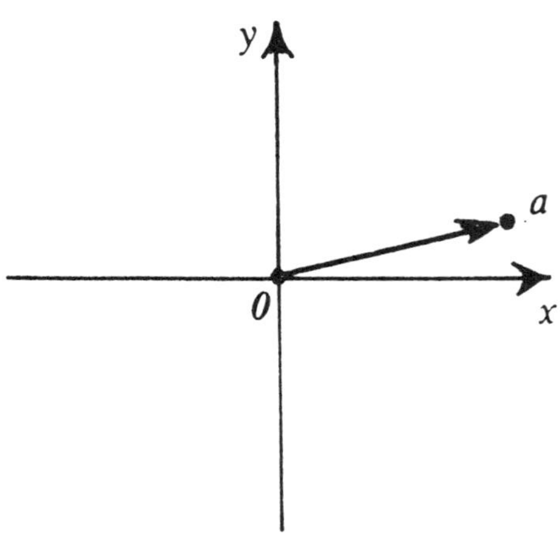

Figure 15.10

By $\boldsymbol{F_1 + F_2}$ we denote the set of points $a_1 + a_2$ where a_1 and a_2 belong to F_1 and F_2 respectively (Figure 15.11).

We leave to the reader to prove the following properties of the vector addition of convex figures.

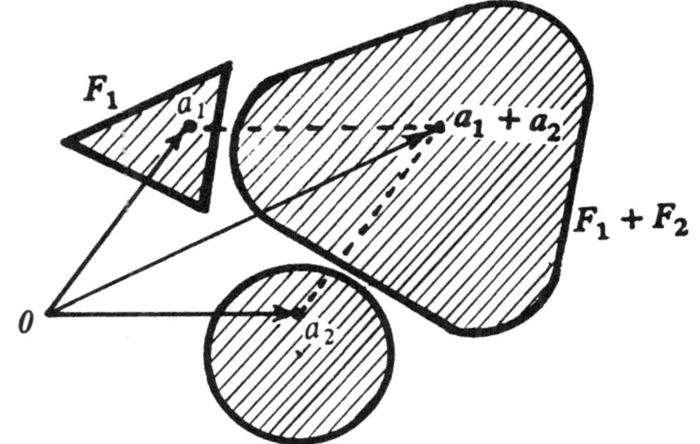

Figure 15.11

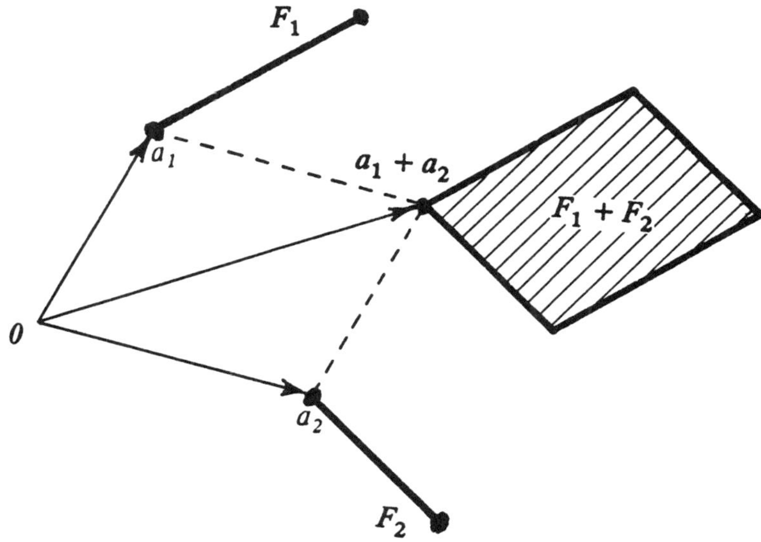

Figure 15.12

For every pair of convex figures F_1 and F_2, the figure $F_1 + F_2$ is also convex.

If F_1 and F_2 are two non-parallel segments, then $F_1 + F_2$ is a parallelogram with sides congruent and parallel to F_1 and F_2 (Figure 15.12).

If L_1 and L_2 are parallel lines (or segments), then $L_1 + L_2$ is a line (segment) that is parallel to L_1 and L_2 (Figure 15.13).

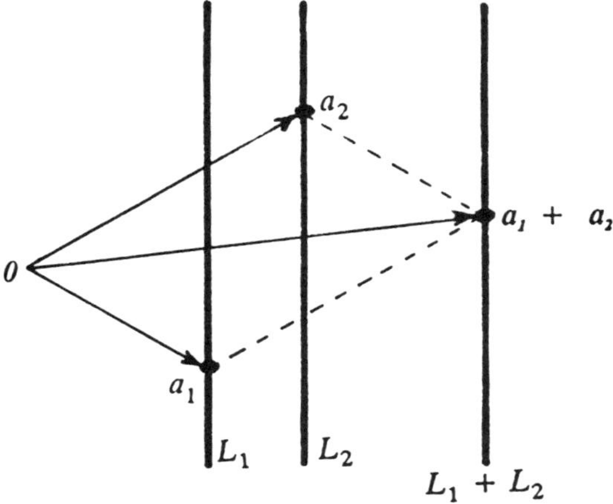

Figure 15.13

Finally, if L_1 and L_2 are parallel support lines of convex figures F_1 and F_2, and F_1 and F_2 are located on the same side of these lines (say "below"), then $L_1 + L_2$ is a support line of the convex figure $F_1 + F_2$ (Figure 15.14).

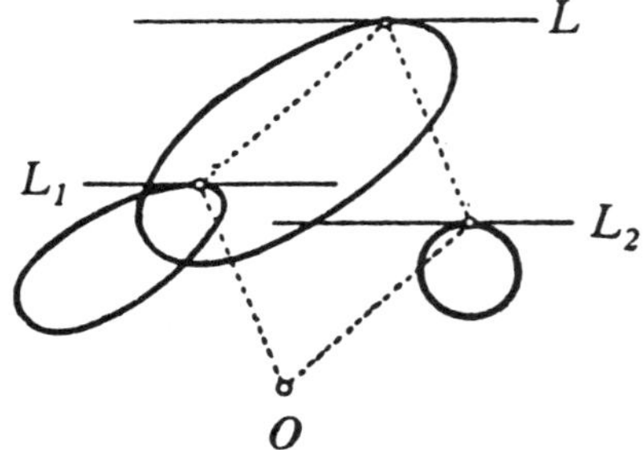

Figure 15.14

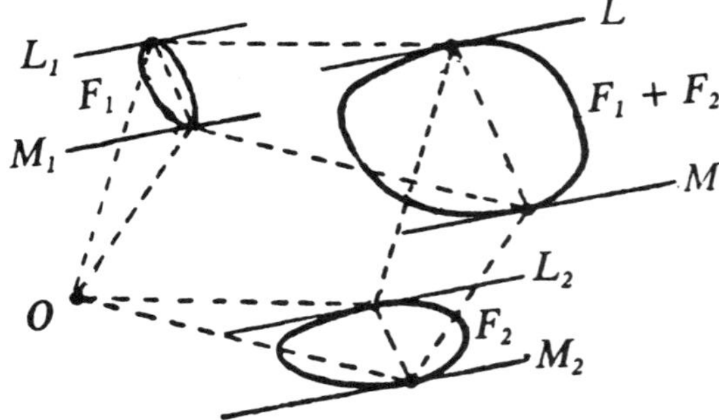

Figure 15.15

This implies that the width of the figure $F_1 + F_2$ in a certain direction is equal to the sum of the widths of the figures F_1 and F_2 in the same direction (Figure 15.15).

Now we can prove the following proposition:

Theorem 15.2. *Let F be a plane figure and F' be the figure that is centrally symmetric to F (with respect to the origin). The figure F has constant width h if and only if the vector sum F + F' is the disk of radius h with center at the origin (Figure 15.16).*

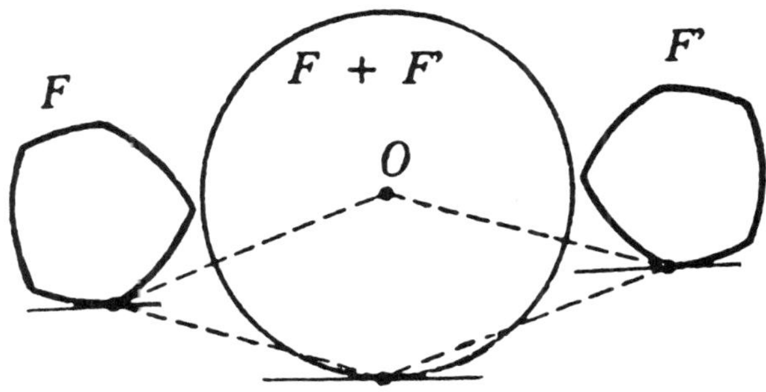

Figure 15.16

Proof. Let F be a figure of constant width h. The width of the figure $F + F'$ in a certain direction is equal to the sum of the widths of the figures F_1 and F_2 in the same direction. Hence, the width of $F + F'$ in any direction is equal to $2h$. Moreover, $F + F'$ is centrally symmetric (with respect to the origin). This means that each support line of $F + F'$ is distance h apart from the origin. Consequently, $F + F'$ is a disk of radius h.

Conversely, let $F + F'$ be a disk of radius h. Then the width of $F + F'$ is equal to $2h$ in any direction. If h_1 is the width of F in a certain direction, then F' has the same width h_1 in this direction, and $F + F'$ has width $2h_1$. Thus, $2h_1 = 2h$, that is, $h_1 = h$. This means that F has width h in any direction, and consequently, F is a figure of constant width h. We are done.

We invite the reader to deduce the Barbier theorem from the above theorem. Another consequence of Theorem 15.2 is the following result:

Inscribed and circumscribed circles of an arbitrary figure of constant width h are concentric and the sum of their radii is equal to h (Figure 15.17).

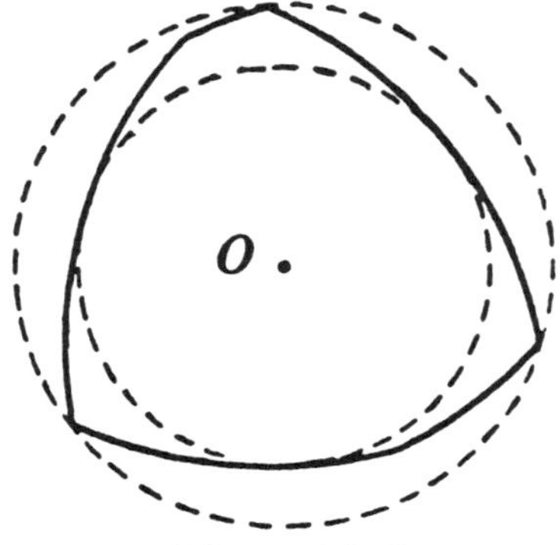

Figure 15.17

Finally, we will formulate one additional result. In Figures 15.3, 15.4, and 15.5 some examples of constant width figures are drawn that are formed by several arcs of circles. In general, such figures can be constructed by the following method. We take a polygon with an odd number of vertices that has the following property: for every vertex there are two diagonals of length h emanating from that vertex, whereas all other diagonals are shorter than h (Figure 15.18). Now we construct all disks of radius h with centers at the vertices of this polygon. Then the intersection of all these disks is a figure of constant width h (Figure 15.19) that is bounded by several arcs of radius h.

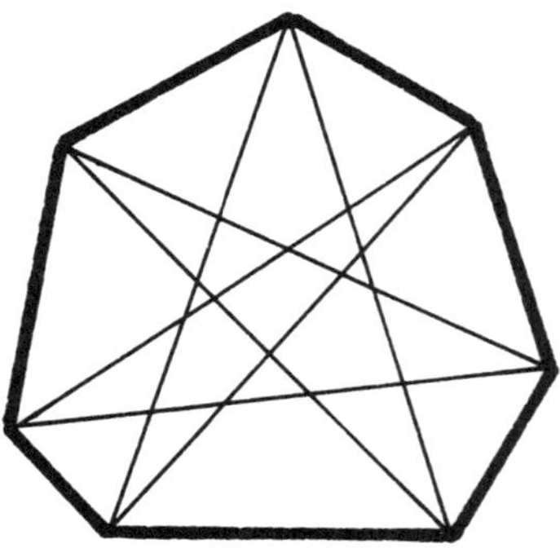

Figure 15.18

It turns out that *every figure of constant width h can be represented as the limit of a convergent sequence M_1, M_2, ... where each M_k is a figure of constant width h that is bounded by several arcs of radius h* (i.e., each M_k is constructed as shown in Figure 15.19.)

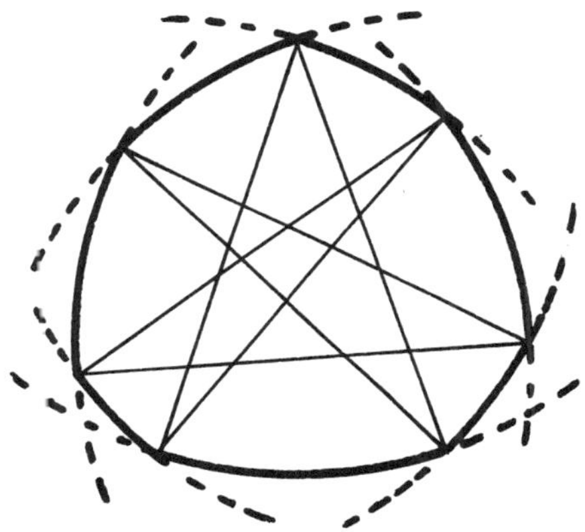

Figure 15.19

Solutions of Exercises

15.1. The tangent line *mc* to the arc *ac* at the point *c* (Figure 15.20) is perpendicular to the chord [*c*,*b*]. Consequently, the angle *mca* is equal to 30°, and the angle *mcn* is equal to 120°. Let L_1 and L_2 be two parallel support lines of the Reuleaux triangle *F*. Let us construct the line *L* parallel to L_1 and L_2 through the

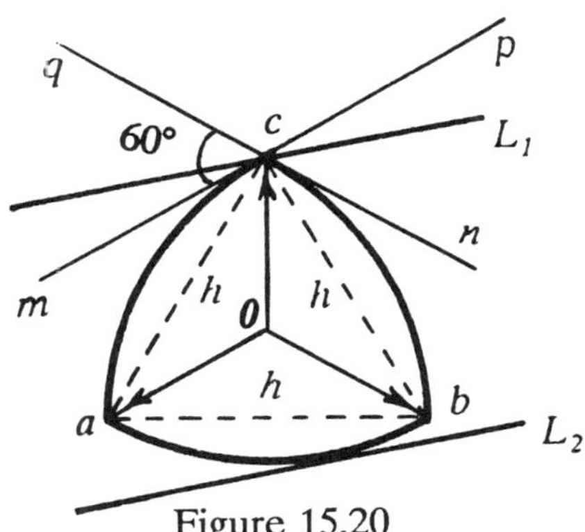

Figure 15.20

176

origin. Then three cases are possible (Figures 15.21, 15.22, and 15.23). In the first case (Figure 15.21) one of the lines L_1, L_2 (say L_1) passes through c, and the other line L_2 is tangent to the arc ab. Hence, the distance between two support lines is equal to the radius of the arc ab, i.e., it is equal to h. The two other cases (Figures 22 and 23) are similar.

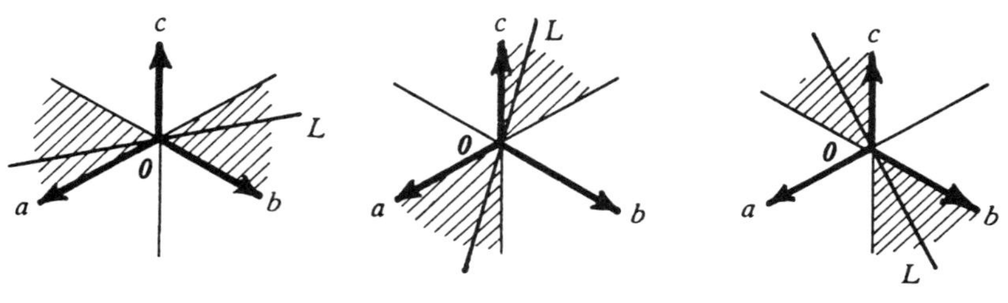

Figure 15.21   Figure 15.22   Figure 15.23

15.2. The solution is similar to the solution of Exercise 15.1 above. If one of two parallel support lines passes through a vertex of M, then the other line is tangent to the opposite arc (Figure 15.24). Therefore, the distance between two parallel support lines is equal to the length of the longest diagonal of M.

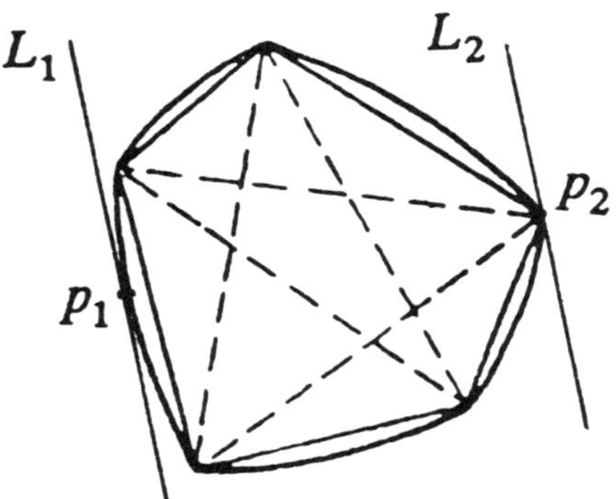

Figure 15.24

15.3. Let a, b be two points of a figure F of constant width h. Further, let L_1, L_2 be two support lines of F that are perpendicular to $[ab]$ with the corresponding support points p_1, p_2, and $p_1 q$

parallel to *ab*. Then $|ab| \leq |p_1 q| = h$ (Figure 15.25). This is true for any pair of points *a,b* of *F*, and consequently, $d \leq h$.

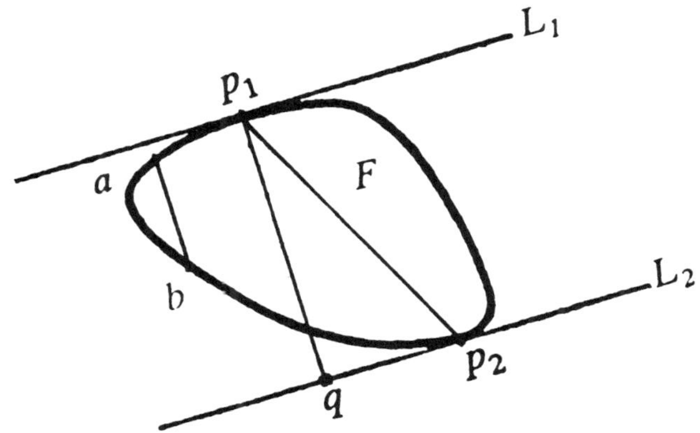

Figure 15.25

Conversely, let L_1, L_2 be two parallel support lines of *F*, and p_1, p_2 be the corresponding support points. Then $|p_1 p_2| \geq |p_1 q| = h$ (Figure 15.25). Hence, $d \geq h$.

This reasoning shows that *q* must coincide with p_2 in Figure 15.25. In other words, if L_1, L_2 are two parallel support lines of a constant width figure and p_1, p_2 are the corresponding support points, then the segment $[p_1, p_2]$ is perpendicular to L_1 and L_2.

15.4. This follows immediately from the solution of Exercise 15.3. Indeed, let *L* be a support line through *a* and *M* the support line parallel to *L* with the corresponding support point *b*. Then $[a,b]$ is perpendicular to *L* and $|ab| = h$ (Figure 15.7). Consequently, $[a,b]$ is a diametral chord of *F*.

Let us note that if *a* is an angular boundary point of *F*, that is, the support line through *a* is not unique, then all the diameters of *F* emanating from *a* form a sector. In other words, the boundary of the figure *F* contains an arc of radius *h* (Figure 15.26).

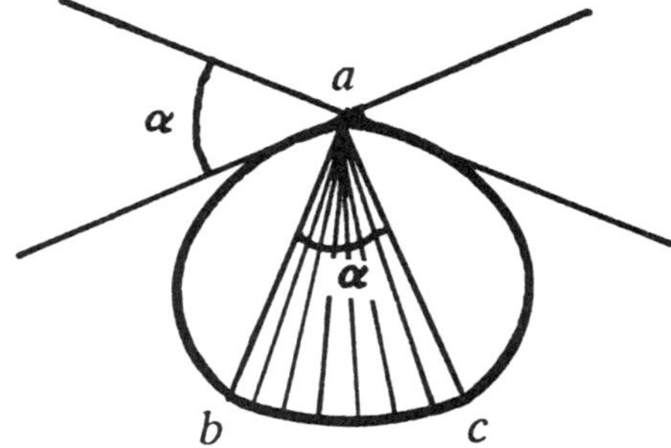

Figure 15.26

It is clear that the length of the chord [b,c] in Figure 15.26 does not exceed h (because the diameter of figure F is equal to h). So, the external angle α at an angular point of the boundary of F does not exceed 60°. It can be easily shown that if the external angle at a corner point of the boundary of F is equal to 60°, then F is the Reuleaux triangle.

15.5. The solution follows immediately from the solution of Exercise 15.3.

15.6. Otherwise we can find points b,c on parallel support lines L *and* M such that [b,c] is not perpendicular to L and consequently $|bc| > h$ (Figure 15.27), contradicting the result of Exercise 15.3.

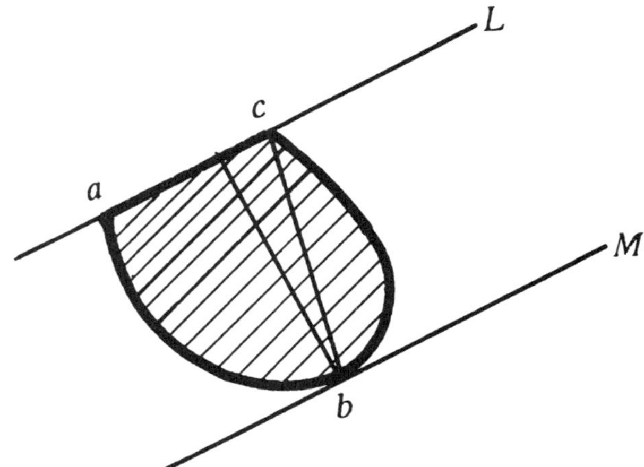

Figure 15.27

15.7. Indeed, if two diametral chords $[a,d]$ and $[b,c]$ have no common points, then we obtain a convex quadrangle $abcd$ inscribed in F with two opposite sides of length h (Figure 15.28). But then at least one of the diagonals of this quadrangle has length greater than h (can you explain why?), contradicting the result of Exercise 15.3.

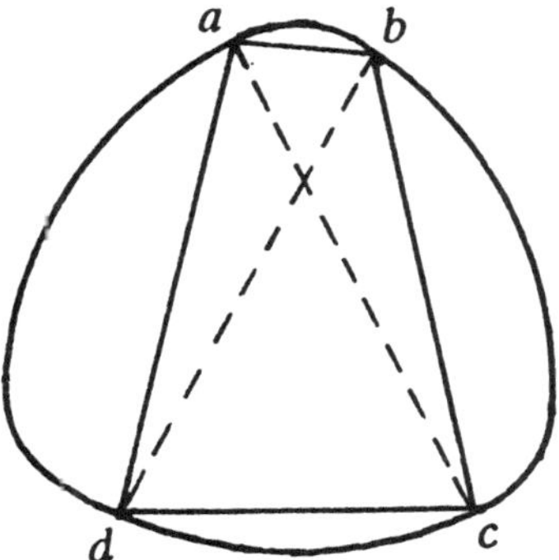

Figure 15.28

15.8. Suppose that Q does not coincide with F, that is, there is a point a in Q that does not belong to F. Then there exists a support line L of the figure F, such that F is located in a closed halfplane P with boundary L, but a does not belong to P (Figure 15.29).

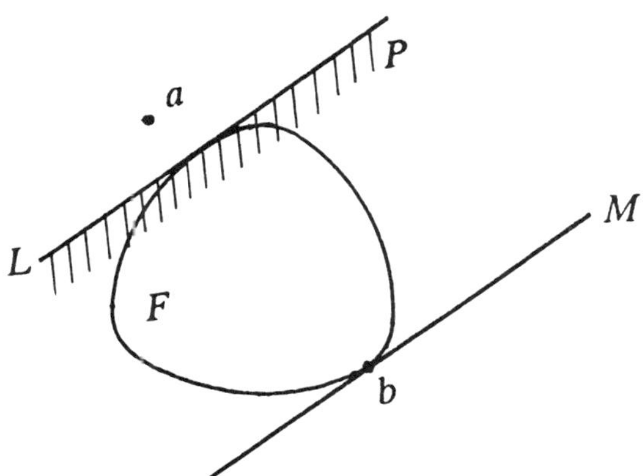

Figure 15.29

Let us denote by M the support line of F parallel to L, and by b the corresponding support point. Then, the distance $|ab|$ is greater than the width of the strip between L and M, that is $|ab| > h$. But both points a and b belong to Q, contradicting the assumption that the diameter of Q is equal to h. This contradiction shows that Q coincides with F.

16. SOLUTION OF THE BORSUK PROBLEM FOR FIGURES IN THE PLANE

In this section we present a solution of the Borsuk problem for the plane, i.e., we describe explicitly in what case a plane figure F satisfies the condition $a(F) = 3$, and in what case $a(F) = 2$ (see page 158 for the definition of the Borsuk number $a(F)$ of a figure F).

First of all we will introduce a new notion and suggest to the reader several exercises.

Let F be a plane figure of diameter not exceeding h. The intersection of all disks of radius h containing F (Figure 16.1) we call the **h-hull** of F and denote by F^*.

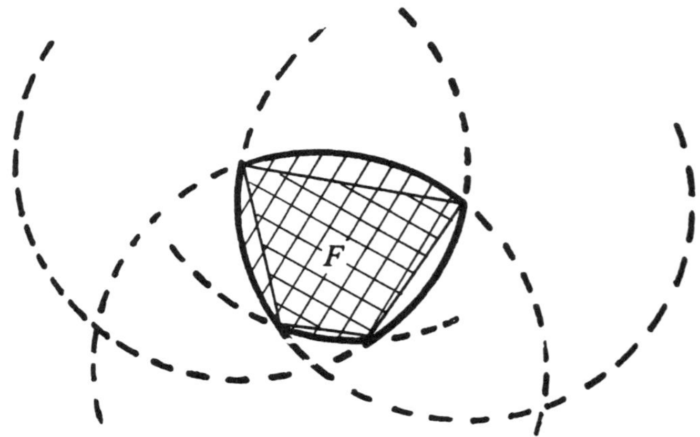

Figure 16.1

Exercise 16.1. Let F be a plane figure of diameter not exceeding h. Prove that the figure F and its h-hull F^* have the same diameter.

Exercise 16.2. Let F be a figure of constant width h. Prove that its h-hull F^* coincides with F.

Exercise 16.3. Let F be a figure of constant width h, L its support line, and a the corresponding support point. Further, let K be the disk of radius h that is tangent to L at the point a and is situated in the halfplane that contains F (Figure 16.2). Prove that F is contained in K.

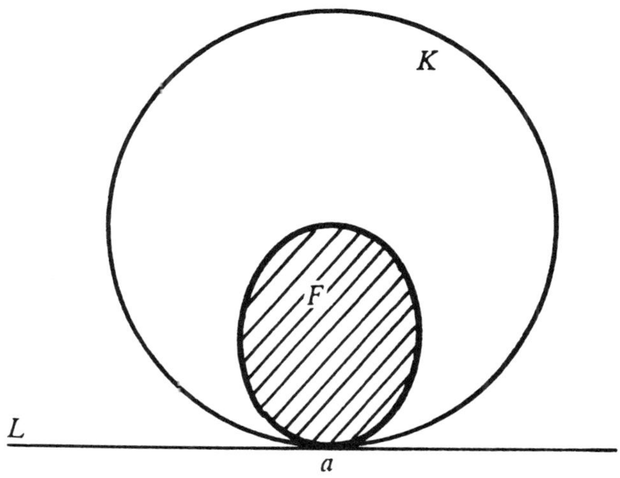

Figure 16.2

Exercise 16.4. Let F be a plane figure of diameter h and F^* its h-hull. Prove that if the width of F^* in at least one direction is less than h, then $a(F) = 2$.

Exercise 16.5. Let F be a plane figure of diameter h such that its h-hull F^* is a figure of constant width h. Let, further, L be a support line for F^* with the corresponding support point a and K be the disk as in Exercise 16.3 (Figure 16.2).

The center q of K is the support point corresponding the support line M for F^* parallel to L (Can you see how this follows from Exercises 15.4 and 15.5?). Denote by b the intersection of M and the circumference of K (Figure 16.3). Prove that if a does not belong to F, then the arc ab contains a point x of F.

Now we are going to prove an important theorem about bounded figures in the plane.

Theorem 16.1. *For each plane figure X of diameter h there exists a figure F of constant width h that contains X.*

Proof. Let us consider the family of all convex figures of diameter h that contain X. By S we denote the exact upper bound of the areas of these figures. Then for each positive integer m

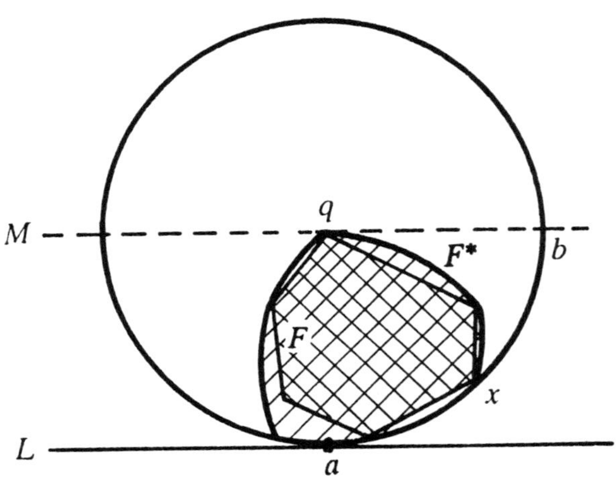

Figure 16.3

there exists in the family a figure F_m, such that its area $S(F_m)$ is greater than $S - \frac{1}{m}$. Each figure F_m contains X and has diameter h. Consequently, all the figures F_m are situated in the disk of radius h with center at a fixed point c of the figure X. By virtue of the Blaschke Theorem (Theorem 13.2), the sequence F_1, F_2, ..., F_n,... contains a converging subsequence. Let F be the limit of this subsequence. It is clear that $S(F) = S$; that is, F has maximal area among figures of diameter h containing X.

Let us prove that F is a figure of constant width h. Evidently the h-hull F^* of the figure F coincides with F (otherwise, F^* would be a greater area figure of diameter h containing X, which is impossible). Assume that $F = F^*$ is not a figure of constant width h. Then there exists two parallel support lines L,M of F with the distance between them less than h. Let a,b be the corresponding support points (Figure 16.4), K be the disk of radius h with center q as in Exercise 16.3. Then q does not belong to F. Taking the convex hull of the union of the point q and the figure F, we obtain a convex figure G of diameter h that contains F and has a greater area, contradicting the choice of F. This contradiction proves that F is a figure of constant width h. This completes the proof.

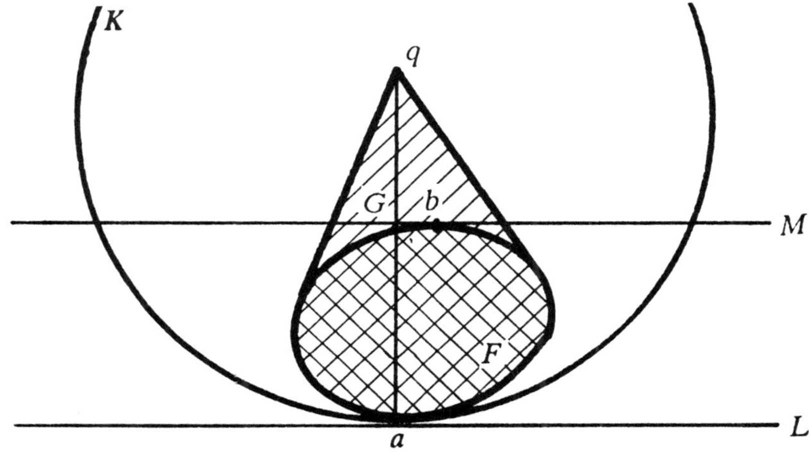

Figure 16.4

Exercise 16.6. Let F be a plane figure of diameter h. Prove that if its h-hull F^* is not a figure of constant width, then there exists at least two distinct figures of constant width h each containing F.

Finally, let us consider a complete solution of Borsuk's problem in the plane found in [B2]. According to Theorem 16.1, each figure X of diameter h is contained in a figure F of constant width h. Is this figure F unique for the given X? Simple examples show that the answer is sometimes "yes" and sometimes "no." For example, if X is an equilateral triangle with the side length h, then there exists only one figure of constant width h containing X, namely, the Reuleaux triangle. Other examples of the same kind we have in Exercise 15.2. Let now X be a segment of a disk of diameter h (i.e., the intersection of the disk and a halfplane) that is larger than the halfdisk (Figure 16.5). Then there are many different figures of constant width h containing X. One of them is the disk of diameter h. The second one is shown in Figure 16.5 (dotted segments have length h). Now we are ready to prove the main result.

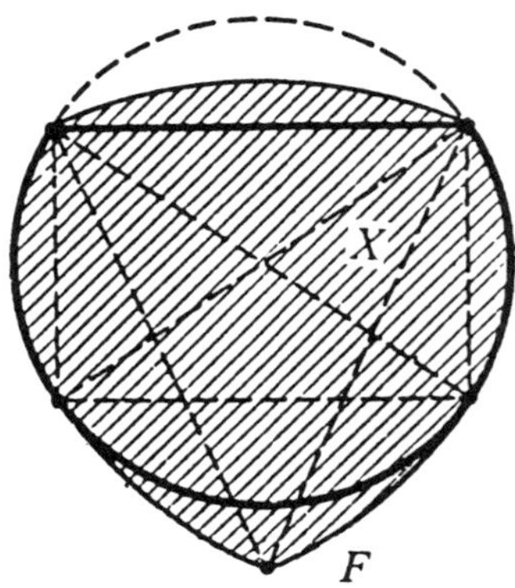

Figure 16.5

Theorem 16.2 (V. Boltyanski [B2]). *Let X be a figure of diameter h in the plane. Then $a(X) = 3$ if and only if the figure of constant width h containing X is unique.*

Proof. Assume that there exists only one figure F of constant width h that contains X. We must prove that $a(X) = 3$. Assume the opposite, i.e., $a(X) = 2$. This means that X is a union of two figures Q_1 and Q_2 each of which has a diameter less than h. Consider the h-hulls X^*, Q_1^*, Q_2^* of the figures X, Q_1, Q_2. According to Exercise 16.6, X^* is a figure of constant width h, that is, X^* coincides with F. Let us prove that the boundary of F is contained in the union of figures Q_1^* and Q_2^*.

Indeed, let a be a boundary point of the figure F. We construct a support line L for F through a and the other support line M for F that is parallel to L (Figure 16.6), and denote by b the corresponding support point. Then due to Exercise 15.5 the segment $[a,b]$ is perpendicular to L and M. We have two possibilities: i) the point a belongs to X; ii) the point a does not belong to X.

In the case i) the point a belongs to Q_1 or Q_2, and consequently, a belongs to Q_1^* or Q_2^*.

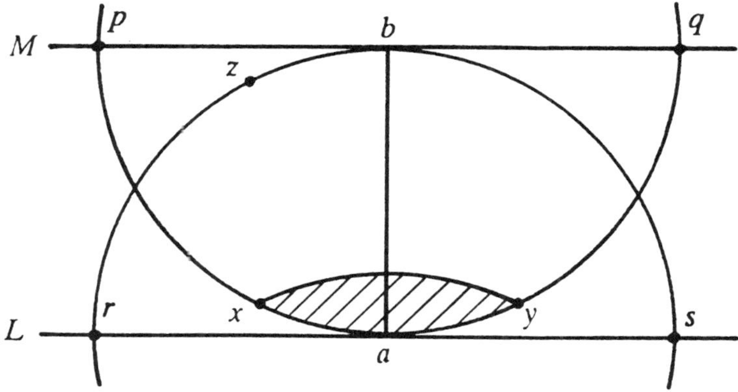

Figure 16.6

Let us consider the case ii). According to Exercise 16.5, the arc ap in Figure 16.6 contains a point x of X. Similarly, the arc aq contains a point y of X. If now the point b does not belong to X, then, clearly, the arc br contains a point z of the figure F. But the distance between y and z is greater than h (can you explain why?), contradicting the assumption that X has diameter h. Consequently, b belongs to X, that is, b is contained in Q_1 or Q_2 (say in Q_1). Further, since both distances $|bx|$ and $|by|$ are equal to h, both points x,y belong to Q_2 (because the diameter of Q_1 is smaller than h). But the lens bounded by two arcs of radius h passing through x and y (it is shaded in Figure 16.6) is contained in *each* disk of radius h that includes x and y. So, the figure $Q_2{}^*$, that is the intersection of all disks of radius h containing Q_2, contains this lens. This means that the point a belongs to $Q_2{}^*$. Thus, in both cases i) and ii), the point a of bdF is contained in the union of the figures $Q_1{}^*$ and $Q_2{}^*$. In other words, bdF can be decomposed into two figures of diameters smaller than h (due to Exercise 16.1, the diameters of $Q_1{}^*$ and $Q_2{}^*$ are equal to the diameters of Q_1 and Q_2 respectively). But this is impossible (that can be proved by the same method, as in the solution of Exercise 14.1). This contradiction shows that the equality $a(X) = 2$ is impossible, that is, $a(X) = 3$.

Conversely, assume that there exist at least two distinct figures F_1, F_2 of constant width h that contain X. Then the intersection ϕ of F_1 and F_2 is not a figure of constant width h.

Moreover, ϕ contains X, and ϕ is the intersection of a family of disks of radius h containing X (since due to Exercise 16.3, F_1, as well as F_2, is the intersection of some family of such disks). So, X^* is contained in ϕ, and consequently, X^* is not a figure of constant width h. Therefore, by virtue of Exercise 16.4, we have $a(X) = 2$. This completes the proof.

We would like to bring your attention to another solution of the Borsuk problem for plane figures that was found by the Polish mathematician Krzysztof Kolodziejczyk [Ko]. In this very interesting solution obtained in 1990, the author does not use figures of constant width.

Let X be a compact convex figure of diameter h and $[c,d]$ be its chord. Kolodziejczyk calls $[c,d]$ a **diametral chord** if $|cd| = h$, and a **non-diametral chord** otherwise.

Theorem 16.3. [Ko] *A compact convex figure X in the plane satisfies the condition $a(X) = 2$ if and only if there exists a non-diametral chord $[c,d]$ of X, such that the endpoints of each diametral chord are situated on different sides of the line cd (Figure 16.7).*

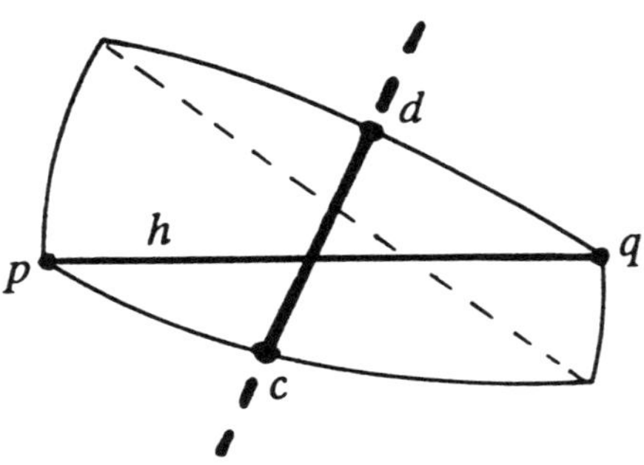

Figure 16.7

For example, the equilateral triangle T whose side has length h does not satisfy the indicated condition. Indeed, for every non-diametral chord $[c,d]$ we can find a diametral chord situated in one closed halfplane with the boundary line cd (Figures 16.8 and 16.9).

Exercise 16.7. Using Theorem 16.2, prove the Kolodziejczyk Theorem.

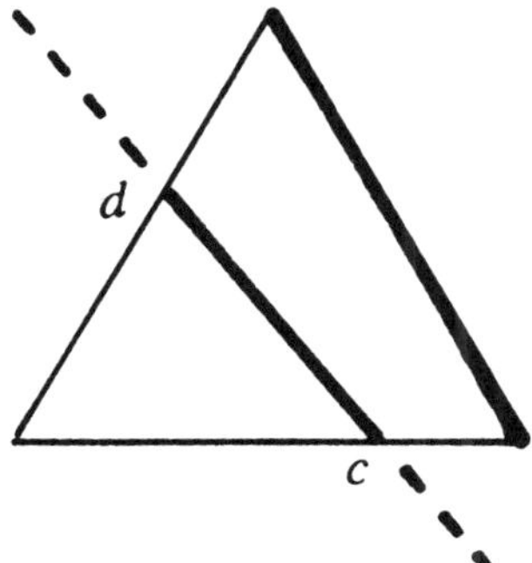

Figure 16.8

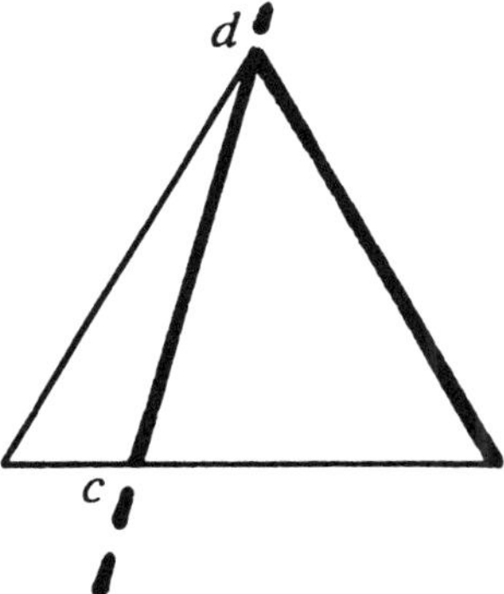

Figure 16.9

Solutions of Exercises

16.1. Assume at first that the diameter d of F is equal to h. Let, further, x be a point of F^* and a a point of F. The circle K_a with center a and radius d contains the whole figure F (because $|ap| \leq d$ for every point p of F). Thus, K_a is one of the disks that have the figure F^* as their intersection. This means that F^* is contained in K_a, and consequently, $|ax| \leq h$. This inequality is true for each point a of the figure F. Hence, the whole figure F is contained in the disk K_x with center x and radius h. So, K_x is one of the disks that have F^* as their intersection. Therefore, F^* is contained in K_x. Thus, $|xy| \leq h$ for every point y of F^*, and consequently, the diameter d_1 of F^* does not exceed h. On the other hand, it is clear that $d_1 \geq h$ since F^* contains F. This completes the proof in this case.

Let us assume now that the diameter d of the figure F is less than h. We denote by F^{**} the d-hull of the figure F. In view of the above, the diameter of the figure F^{**} is equal to d. It can be easily shown that the h-hull F^* of F is contained in F^{**}. Indeed, if a point z does not belong to F^{**}, then there exists a disk Q of radius d, such that Q contains F, and z does not belong to Q. Hence, there exists a disk K of radius h that contains Q and does not contain z (since $d < h$; see Figure 16.10). Then K is one of the disks that have F^* as their intersection, and consequently F^* is contained in K. This means that z does not belong to F^*.

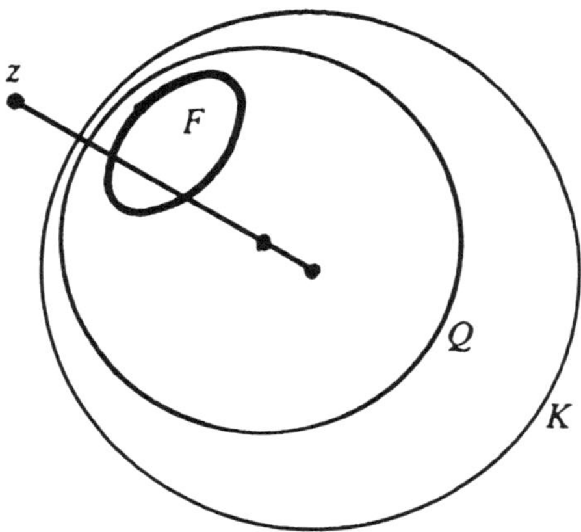

Figure 16.10

Thus, if z is not contained in F^{**} then z does not belong to F^*, that is, F^* is contained in F^{**}. Hence the diameter d_1 of F^* does not exceed d. On the other hand, $d_1 \geq d$ (since F^* contains F). This completes the proof.

16.2. The diameter of the figure F is equal to h, and consequently, the diameter of F^* is equal to h as well (Exercise 16.1). Now the validity of the assertion of Exercise 16.2 follows immediately from Exercise 15.8 of the previous section.

16.3. *Let us prove a more general proposition from which Exercise 16.3 follows by virtue of Exercise 16.2. Let F be a figure of diameter h and F* its h-hull. Further, let L be a support line of F*, a the corresponding support point, and K the disk of radius h that is tangent to L at a and is situated in the halfplane P that contains F*. Then F* is contained in K.*

Indeed, assume that there exists a point x in F^* that does not belong to K (Figure 16.11). Denote by M the "lens" bounded by two arcs of radius h passing through the points a and x. Each disk of radius h containing the points a,x contains the whole lens M. Consequently, this lens is contained in F^* (since both points a and x belong to F^*, and F^* is the intersection of a family of disks of radius h). But the lens M is not contained in the halfplane P, contradicting the assumption that L is a support line of F^*. This contradiction proves the proposition.

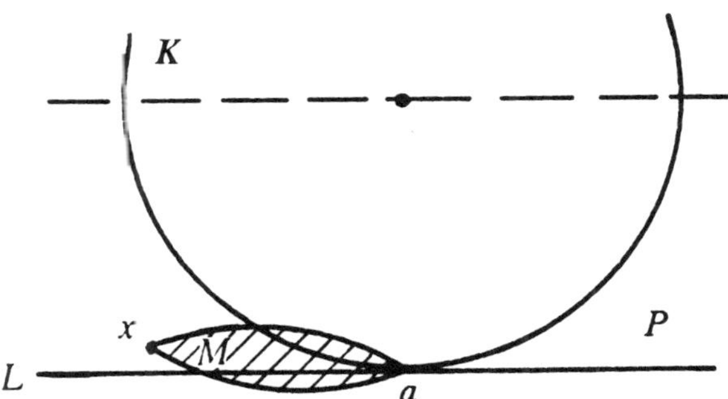

Figure 16.11

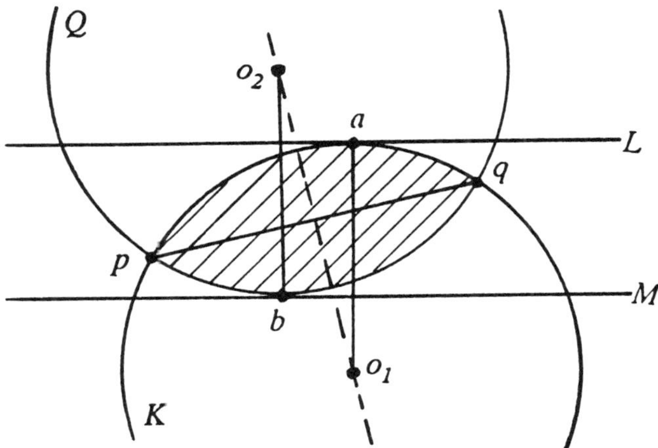

Figure 16.12

16.4. Let L and M be two parallel support lines of F^*, such that the distance between them is less than h (Figure 16.12), and a and b be the corresponding support points. Then F^* is contained in the disk K of radius h that is tangent to L at the point a. Similarly, F^* is contained in the disk Q of radius h that is tangent to M at point b. The intersection of the disks K and Q is a "lens" bounded by two arcs of radius h. The centers of the disks K and Q are situated outside of the lens, and consequently, the distance between the endpoints p and q of the lens is less than $2h$. Therefore the line passing through the centers of the disks decomposes the lens (and the figure F^* contained in the lens) into two parts of diameter less than h.

16.5. If the arc ab does not contain any point of F, then for small enough $\mu > 0$ the image K^1 of K under the translation through the vector $\mu \overrightarrow{bq}$ contains F (Figure 16.13). Consequently, K^1 is one of the disks that have F as their intersection, and, hence, F^* is contained in K^1. Therefore, F^* is contained in the intersection of K and K^1, contradicting the assumption that L is a support line of F^*.

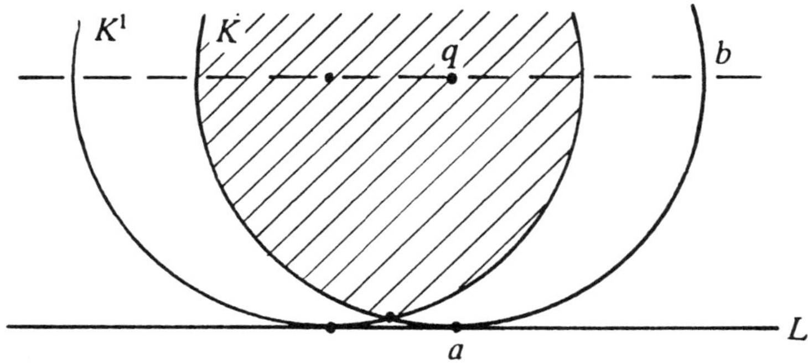

Figure 16.13

16.6. Let us repeat the construction discussed in the solution of Exercise 16.4 (see Figure 16.12). The convex hull of the union F_1 of the figure F and the center o_1 of the disk K is a figure of diameter h (indeed, the distance between o_1 and any point x of the figure F does not exceed h since F is contained in the lens). Similarly, the convex hull of the union F_2 of the figure F and the center o_2 of the disk Q is a figure of diameter h. Consequently, by virtue of Theorem 16.1, F_1 is contained in a figure H_1 of constant width h, and F_2 is contained in a figure H_2 of constant width h. Finally, notice that H_1 does not coincide with H_2, since the distance $|o_1o_2|$ exceeds h (indeed, $|o_1o_2| > |o_1a| = h$).

16.7. If there exists a non-diametral chord $[c,d]$, as in Theorem 16.3, then, evidently, the line cd divides X into two parts of smaller diameters, that is, $a(X) = 2$.

Conversely, let $a(X) = 2$. Then, by virtue of Theorem 16.2, there are two distinct figures of constant width h that contain X. Hence the h-hull X^* is not a figure of constant width h (see the end of the proof of Theorem 16.2). This means that the width of X^* in a certain direction is less than h. Consequently (see the solution of Exercise 16.4), the h-hull X^* is contained in a lens with angle points p and q, such that $|pq| < h\sqrt{3}$ (see notations in Figure 16.12). This means that the endpoints of each diametral chord of X^* (and of X too) are situated on different sides of the line o_1o_2. Thus, the line o_1o_2 intersects X in the required non-diametral chord $[c,d]$.

17. ILLUMINATION OF CONVEX FIGURES

In Figure 17.1 you see a convex figure F. A parallel beam of light illuminates a part of its boundary. The points a and b are not illuminated. In other words, a boundary point c is ***illuminated***, if for small enough $\mu > 0$ the point c' for which $\overrightarrow{cc'} = \mu\overrightarrow{w}$ belongs to the interior of F, where $\overrightarrow{w}$ is a defining vector of the beam.

The ***illumination problem*** asks for the minimal number of such parallel beams that illuminate the whole boundary of the given convex figure F. We denote this minimal number by $c(F)$. Figure 17.2 shows that for a disk F we have $c(F) = 3$. And if F is

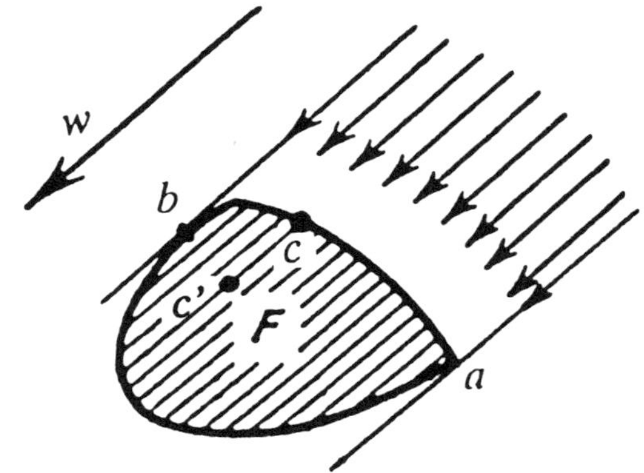

Figure 17.1

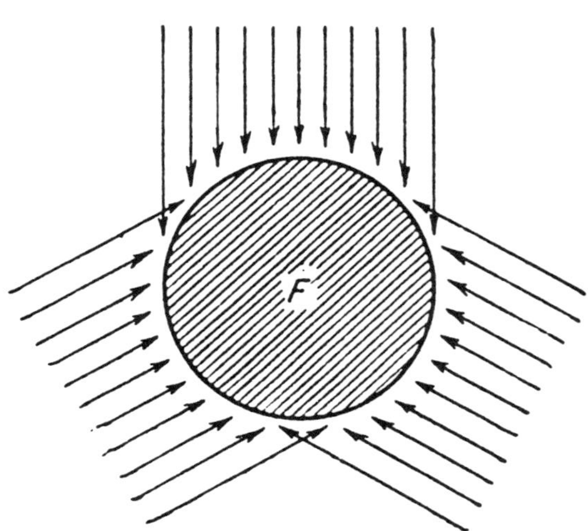

Figure 17.2

a parallelogram, then $c(F) = 4$. Indeed, two vertices cannot be illuminated by the same beam, that is, each vertex needs its own beam (Figure 17.3). It can be easily shown that if a convex quadrangle F is not a parallelogram (for example, if F is a trapezoid), then $c(F) = 3$ (Figure 17.4).

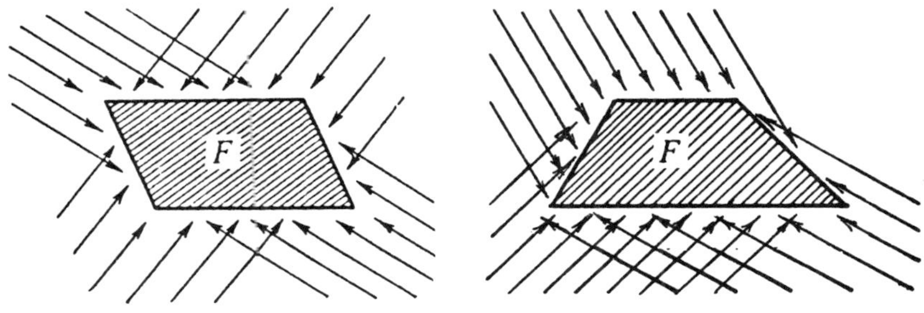

Figure 17.3   Figure 17.4

The illumination problem was born in 1960, in the paper [B3]. It makes sense not only for plane figures, but also for solid bodies (and even for n-dimensional bodies, $n > 3$). A complete solution of the problem for plane figures is contained in the following proposition:

Theorem 17.1. *If a convex figure F in the plane with a non-empty interior is not a parallelogram, then $c(F) = 3$. For each parallelogram P, we get $c(P) = 4$.*

Can you find a proof of this theorem? If not, then solve the following exercises that lead to a proof of Theorem 17.1.

Exercise 17.1. Prove that if the boundary of a plane convex figure F does not contain angle points (that is, all the boundary points of F are regular), then $c(F) = 3$.

Exercise 17.2. Let F be a plane convex figure and p an angle point of its boundary. Prove that there exists a parallelogram *pqrs* circumscribed about F, such that the rays emanating from p and passing through q and s respectively are tangent to F (Figure 17.5).

Exercise 17.3. Let F be a plane convex figure, p an angle point of its boundary and *pqrs* the circumscribed parallelogram, as in Exercise 17.2. Prove that if the point r does not belong to F (Figure 17.6), then $c(F) = 3$.

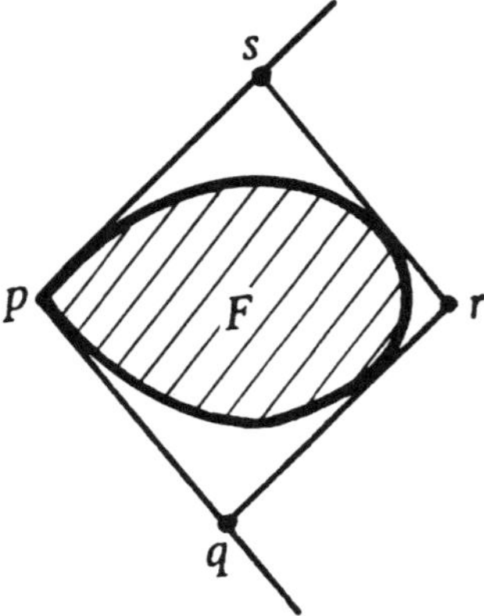

Figure 17.5

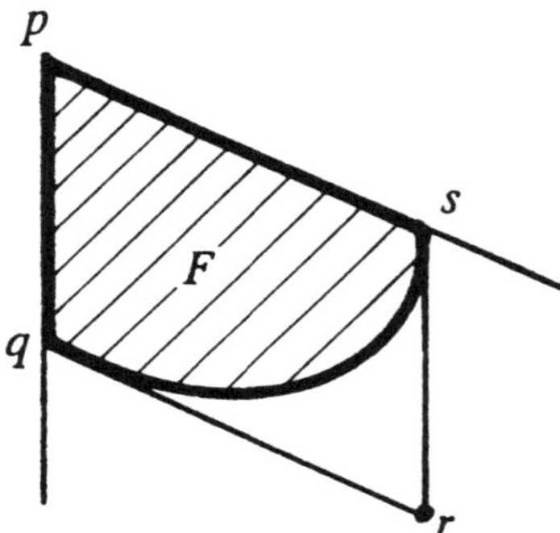

Figure 17.6

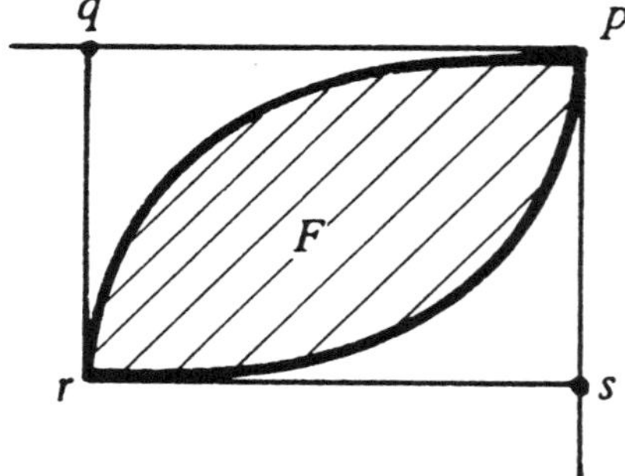

Figure 17.7

Exercise 17.4. Let F be a plane convex figure and p an angle point of its boundary and $pqrs$ the circumscribed parallelogram, as in Exercise 17.2. Prove that if at least one of the points q,s does not belong to F (Figure 17.7), then $c(F) = 3$.

Exercise 17.5. Give a proof of Theorem 17.1.

We are done with Flatland. How are things in space? It is clear that if F is a parallelepiped, then $c(F) = 8$. Indeed, no two vertices of F can be illuminated by one light beam, that is,

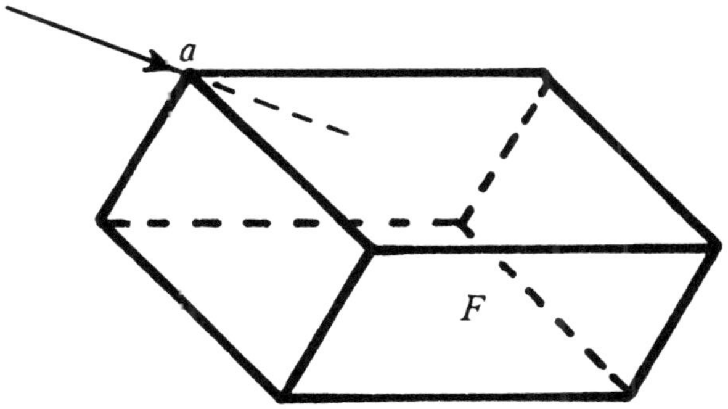

Figure 17.8

each vertex needs its own beam (Figure 17.8). Further, in the plane each parallelogram was a figure with the *maximal* value of $c(F)$. Does the similar proposition hold for the space? In other words,

is it true that for all convex bodies F except parallelepipeds the inequality $c(F) < 8$ holds?

This is still an open problem! Even for polyhedrons nobody knows the answer. Only for centrally symmetric convex bodies is a solution known: the Polish mathematician M. Lassak established a positive answer for such bodies. In n-dimensional space R^n this question is known as **The Hadwiger Problem:**

Is it true that for all convex bodies F in R^n except parallelotopes (i.e., n-dimensional analogy of parallelepipeds) the inequality $c(F) < 2^n$ holds?

In case $n > 4$ the answer is unknown even for centrally symmetric polytopes. But in several partial cases a solution of the Hadwiger problem is known.

Exercise 17.6. Prove that if a convex body F in R^3 is smooth (that is, each boundary point of F is regular), then $c(F) = 4$.

A similar result was proved in 1945 by the Swiss mathematician H. Hadwiger in [H1] for n-dimensional space R^n for all positive integers n: $c(F) = n + 1$ for all smooth convex bodies F in R^n.

Moreover, if a convex body F in R^n has at most n angle points, then the equality $c(F) = n + 1$ is true (see [B2]).

In R^3 a stronger result is known (see [K]): if a convex body F in R^3 has no more than 4 angle points, then $c(F) = 4$.

There is a very interesting relationship between the illumination problem and another combinatorial problem posed by Hugo Hadwiger in [H3].

Let F be a figure. Choose an arbitrary point q in the plane, and, in addition, choose a positive number k. For any point p of the figure F we shall find a point p' on the ray qp such that $|qp'| \div |qp| = k$ (Figure 17.9). The set of all points so obtained is represented by a new figure F'. The transition from the figure F to the figure F' is called *homothety* with center q and coefficient k, and the figure F' itself is called a *homothetic image* of F. (Homothety with a negative coefficient will not be necessary for us in what follows, and we shall therefore not consider it.)

If a figure F is convex, then its homothetic image F' is also convex (can you prove it?).

We will call a figure F_1 a *smaller copy* of F, if F_1 is homothetic to F with a center q and a positive ratio $k_1 < 1$ (Figure 17.9).

The Hadwiger Covering Problem asks

What is the minimal number n such that n smaller copies of a convex figure F can cover the whole figure F?

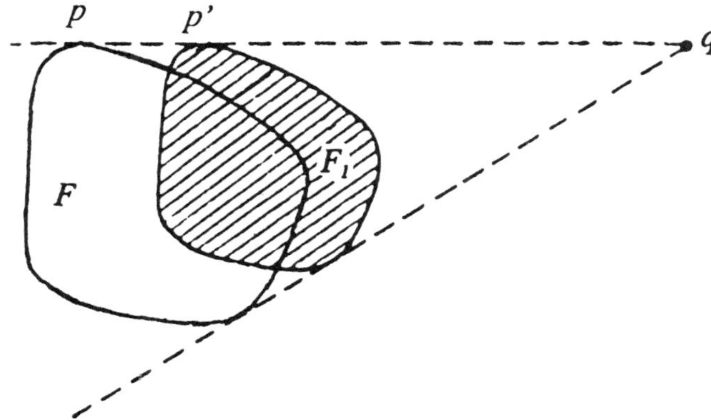

Figure 17.9

This problem can be considered in the plane, as well as in the space (and in *n*-dimensional space R^n). We denote the minimal number of smaller copies covering F by $b(F)$. H. Hadwiger found a solution of the problem for plane figures in [H1]. He also proved that *for smooth convex bodies in R^n the equality $b(F) = n + 1$ is true.*

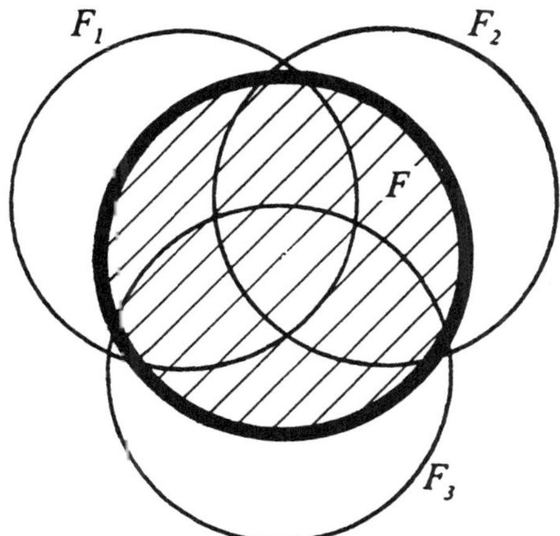

Figure 17.10

For example, if F is a disk, then $b(F) = 3$ (Figure 17.10). If F is a parallelogram, then $b(F) = 4$. Indeed, no two vertices of a parallelogram can be covered by its smaller copy (Figure 17.11). Thus, each vertex needs its own smaller copy.

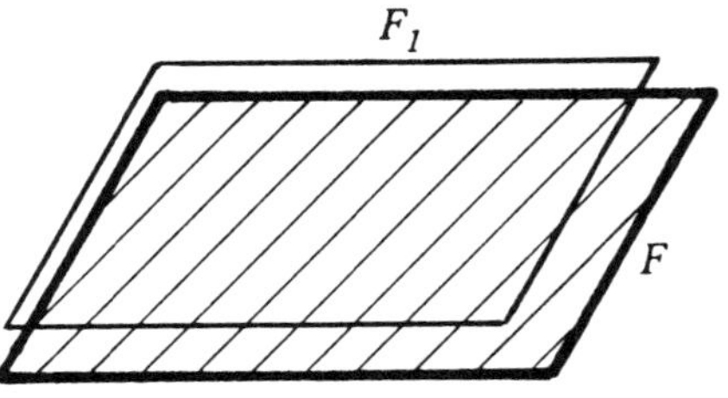

Figure 17.11

The events in the illumination problem were very similar, weren't they? Perhaps these two problems have the same answer for every convex figure F? The following theorem explains when it is indeed true.

Theorem 17.2 (Boltyanski, [B1]). *For every compact convex body F in R^n, the equality $b(F) = c(F)$ is true.*

In particular, this is true for compact convex figures in the plane:

Exercise 17.7. Prove that for every convex plane figure F that is not a parallelogram, the equality $b(F) = 3$ is true.

The next exercise shows that for unbounded figures the assertion of Theorem 17.2 is in general false.

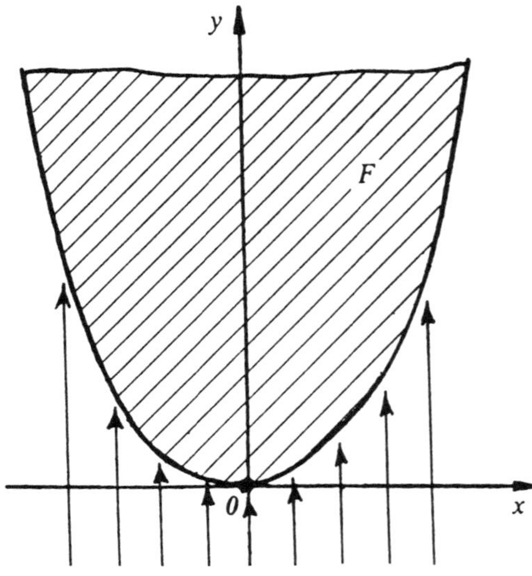

Figure 17.12

Exercise 17.8. The figure F is defined in the cartesian coordinate system by the inequality $y \geq x^2$, that is, F is the convex figure whose boundary coincides with the parabola $y = x^2$. Figure 17.12 shows that $c(F) = 1$. Prove that $b(F) = \infty$.

Finally, let us look into a connection between the problems discussed in this section and the Borsuk problem.

Theorem 17.3. *For every compact convex body F in R^n the inequality $a(F) \leq b(F)$ is true. Thus, we have*

$$a(F) \leq b(F) = c(F)$$

This assertion is trivial. Indeed, if $b(F)=s$, than the body F can be covered by s smaller copies of F. And it is clear that if the diameter of F is equal to d, then the diameter of each of its smaller copies is less that d. This completes the proof.

Finally, due to Exercise 17.6, if a convex body F in R^3 is smooth then $a(F) \leq 4$, in agreement with the Eggleston result of Section 14.

A similar result holds in R^n: For every smooth convex body F in R^n the inequality $a(F) \leq n + 1$ is true.

Moreover, if a convex body F in R^n has at most n non-regular boundary points, then $a(F) \leq n + 1$.

Solutions of Exercises

17.1. Let pqr be a triangle and h its interior point (Figure 17.13). We will show that the three light beams defined by the vectors $\overrightarrow{ph}$, $\overrightarrow{qh}$, and $\overrightarrow{rh}$ illuminate the boundary of each convex figure F whose boundary consists only of regular points.

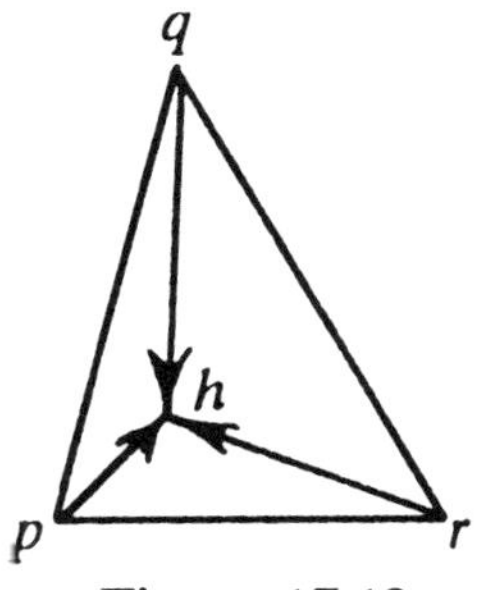

Figure 17.13

Indeed, let x be a boundary point of F. Consider the triangle $p'q'r'$ that is the image of pqr under the translation through the vector $\overrightarrow{hx}$ (Figure 17.14). Let L be the support line of F through x (it is unique because the boundary point x is

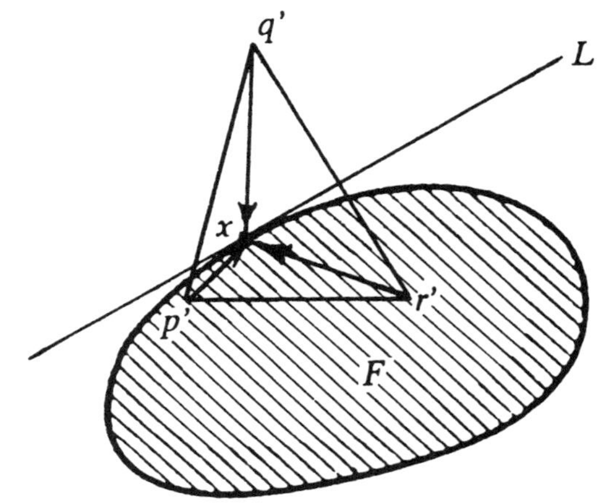

Figure 17.14

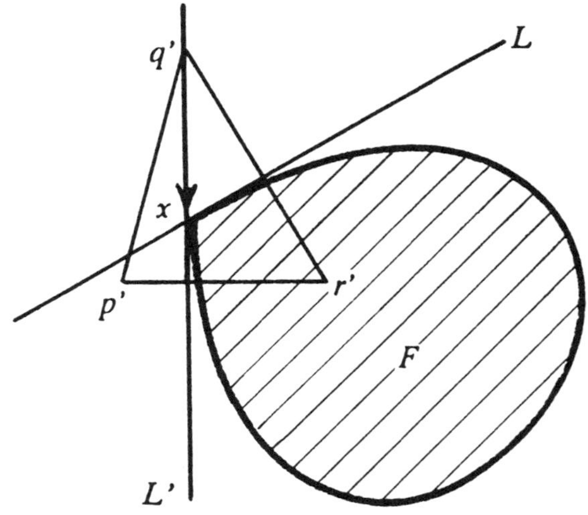

Figure 17.15

regular). Then at least one of the points p', q', and r', say q', is situated on the other side of the line L than the figure F. Now it is clear that the beam of direction $\overrightarrow{q'x} = \overrightarrow{qh}$ illuminates the point x, that is, the line L' through the points q' and x intersects an interior of the figure F. Indeed, otherwise L' and L (Figure 17.15) would be two distinct support lines of F at x, contradicting the assumption that x is a regular boundary point.

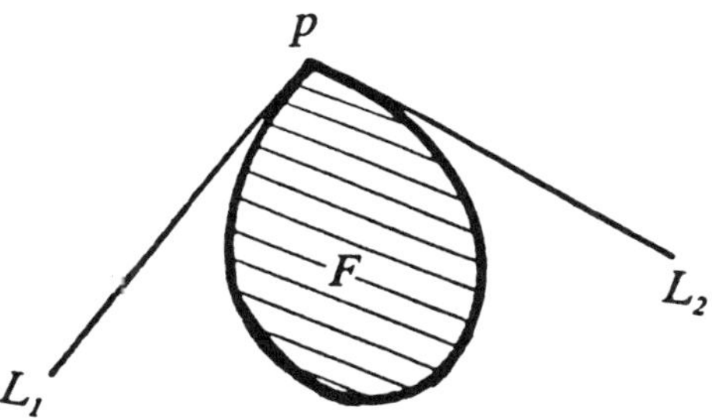

Figure 17.16

17.2. Let L_1 and L_2 be two rays that are tangent to F at the boundary point p (Figure 17.16). Further, let M_1 and M_2 be support lines of F parallel to L_1 and L_2 respectively (Figure 17.17). Then the rays L_1 and L_2 and the lines M_1 and M_2 determine the required parallelogram.

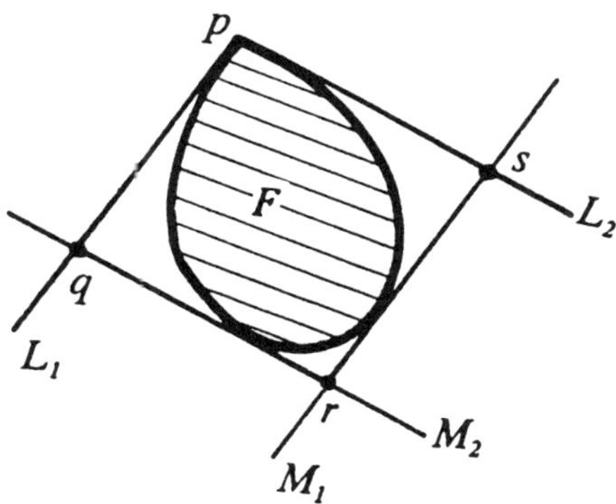

Figure 17.17

17.3. Let x and y be boundary points of F situated in the arc qs that do not belong to the boundary of the parallelogram $pqrs$ (Figure 17.18). Then the beam of direction $\overrightarrow{xq}$ illuminates all the points of the arc $psyx$ except p (Figure 17.18). Similarly, the beam of direction $\overrightarrow{ys}$ illuminates all the point of the arc $pqxy$ except p (Figure 17.19). Thus, these two beams illuminate the whole boundary of F except the point p. It remains to take a third direction that illuminates p.

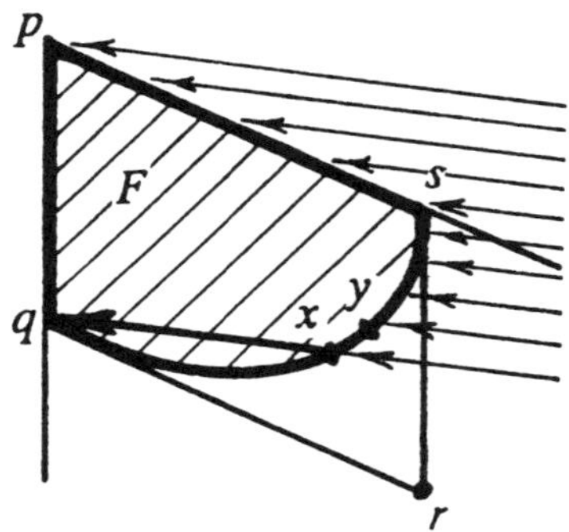

Figure 17.18

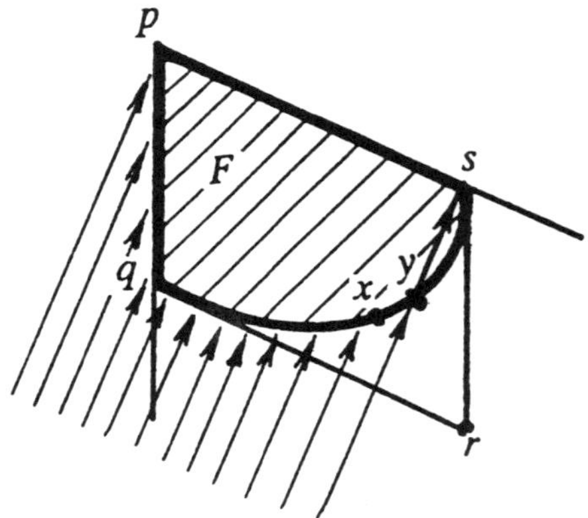

Figure 17.19

17.4. Assume, for definiteness, that the point q does not belong to F. We may also assume that r is contained in F (otherwise the result follows from Exercise 17.3). Further, let m_1 and m_2 be rays emanating from r and tangent to F at the point r. First we consider the case when at least one of the rays m_1, m_2 say m_1, is situated in the interior of the angle qrs (Figure 17.20).

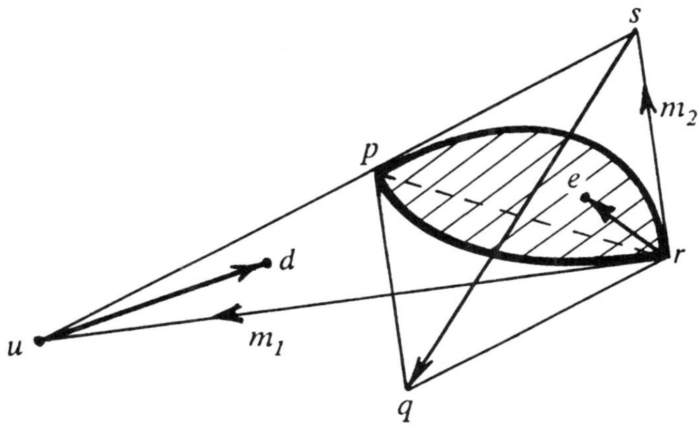

Figure 17.20

Let us denote by u the point of intersection of the line ps and the ray m_1. Further, we choose an interior point d of the angle sur. Then the beam of direction $\overrightarrow{ud}$ illuminates all the points of the arc pr (situated in the triangle pqr) including p, but excluding r. The beam of direction $\overrightarrow{sq}$ illuminates all the points of the arc pr (situated the triangle spr) except endpoints p and r. Finally, if e is an interior point of the angle between the rays m_1 and m_2 then the beam of direction $\overrightarrow{re}$ illuminates the point r. Thus, the whole boundary of F is illuminated by three light beams of directions $\overrightarrow{ud}$, $\overrightarrow{sq}$, and $\overrightarrow{re}$.

It remains to consider the case when rays m_1 and m_2 pass through the points q and s respectively. Let x be a boundary point of F situated in the arc pr (in the triangle pqr) that does not belong to the boundary of the parallelogram $pqrs$ (Figure 17.21). The point x does not belong to the diagonal $[p,r]$ (since the ray emanating from p and passing through q is tangent to F). Let us denote by u the point of intersection of the lines sp and rx, and by v the point of intersection of the lines sr and px. Further, we choose an interior point d of the angle sur and an interior point

e of the angle *svp*. Then the beam of direction $\vec{ud}$ illuminates all the points of the arc *px* (including endpoints) and the beam of

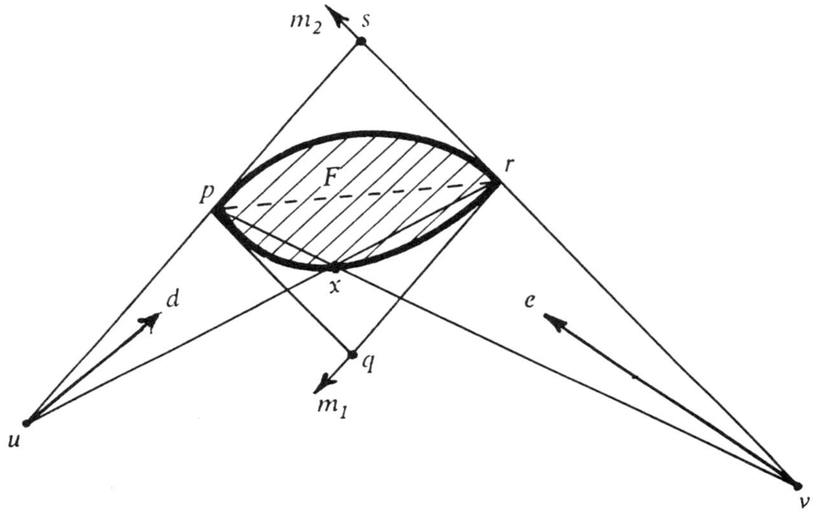

Figure 17.21

direction $\vec{ve}$ illuminates all the points of the arc *rx* (including endpoints). Finally, the beam of direction $\vec{sq}$ illuminates all the points of the arc *pr* (in the triangle *prs*) except endpoints. So, the beams of directions $\vec{ud}$, $\vec{ve}$, and $\vec{sq}$, illuminate the whole boundary of F in this case.

17.5. Exercises 17.1, 17.2, 17.3, and 17.4 give a proof of Theorem 17.1. Indeed, if F is smooth, then $c(F) = 3$ (Exercise 17.1). Let us assume now that the boundary of F contains at least one angle point p. We consider the parallelogram *pqrs*, as in Exercise 17.2 (see Figure 17.5). If r does not belong to F, then $c(F) = 3$ (Exercise 17.3). If r is contained in F, but at least one of the points q or s does not belong to F, then $c(F) = 3$ as well (Exercise 17.4). Finally, when all the points p, q, r, and s are contained in F, F coincides with the parallelogram *pqrs*.

17.6. The reasoning is similar to the solution of Exercise 17.1. We consider a tetrahedron *pqrs* and its interior point h. Then the beams of directions $\vec{ph}$, $\vec{qh}$, $\vec{rh}$, and $\vec{sh}$ illuminate the boundary of each smooth convex body F. Indeed, let $p'q'r's'$ be the image of the tetrahedron $pqrs$ under the translation through the vector $\overrightarrow{hx}$ where x is a boundary point of F. Let L be the support plane of F through x. Then at least one of the points p', q', r', and s', say q', is situated on the other side of the plane L than the body F. Now it is clear that the beam of direction $\overrightarrow{q'x} = \overrightarrow{qh}$ illuminates the point x, that is, the line l' through q' and x intersects the interior of F. Indeed, otherwise there exists a support plane L' containing l'. The plane L' does not coincide with L, since L does not contain l'. Thus, there are two support planes L and L' through x, contradicting the assumption that the boundary point x is regular.

17.7. Let us consider three beams that illuminate the whole boundary of F (Figure 17.22). Then the tangent points b and c are distinct, the tangent points d and e are distinct too (see notations in Figure 17.22), and the tangent points f and a are distinct. We choose the points p, q, and r between these pairs of distinct points respectively and denote by g an interior point of F. Then F is the union of the three figures P_1, P_2, and P_3, each of which is the convex hull of the point g and one of the arcs pq, qr, and rp (the figure P_1 is shaded twice in Figure 17.22). Notice that

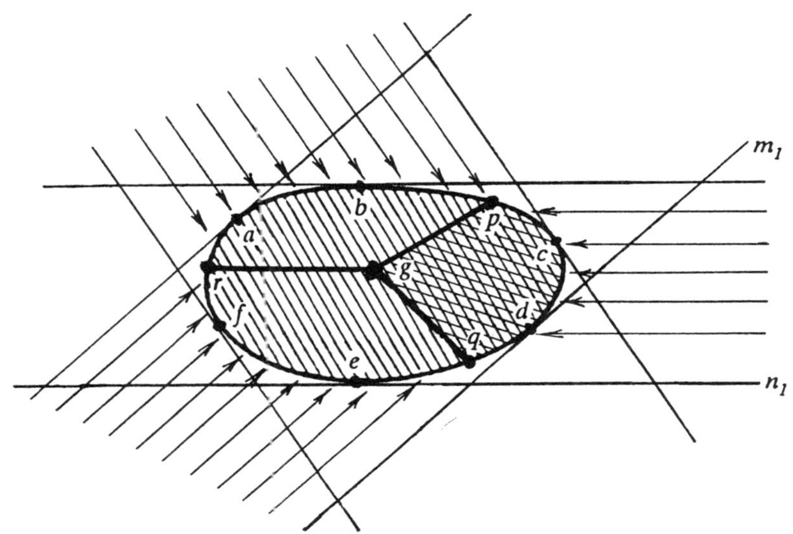

Figure 17.22

P_1 is situated *strictly between* the support lines m_1 and n_1. So if a point o_1 is situated far enough (between m_1 and n_1) and the ratio k_1 is close enough to 1, then the homothety h_1 with the center o_1 and the ratio k_1 maps F onto the figure $h_1(F)$ that contains P_1. Choosing similarly homotheties h_2 and h_3, we obtain the required smaller copies $h_1(F)$, $h_2(F)$, $h_3(F)$ of F.

This reasoning can be generalized to the n-dimensional case.

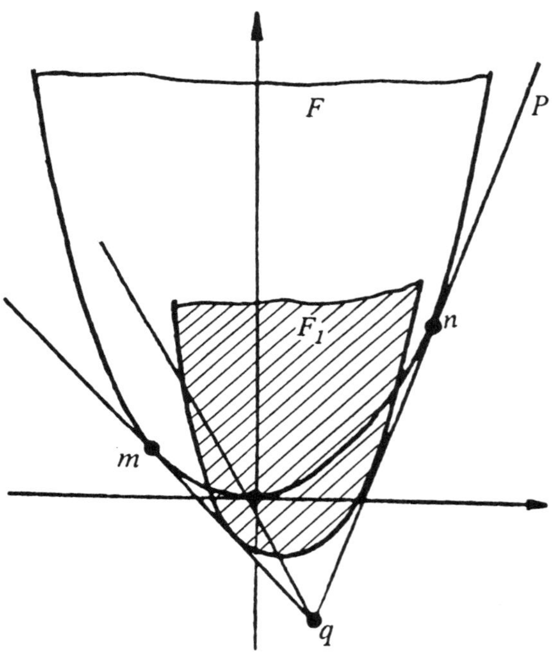

Figure 17.23

17.8. Let q be a point that does not belong to F (Figure 17.23). Then there exist two tangent lines to the parabola P through q. Denote by m,n the corresponding support points. These points bound an arc mn of finite length. If now F_1 is a "smaller copy" of F, that is, the image of F under the homothety with the center q and a positive ratio $k < 1$, then the intersection of F_1 with the parabola P is an arc contained in the arc mn. Thus, each "smaller copy" covers only a finite length arc of P, and, consequently, it takes infinitely many "smaller copies" of F to cover F. This means that $b(F) = \infty$.

18. THEOREMS OF HELLY AND SZÖKEFALVI-NAGY

The first of the theorems to be discussed in this section was discovered by the Austrian mathematician, Eduard Helly, in 1913. But it was not published at that time. During World War I, E. Helly was a soldier of the Austrian Army, and he was taken a Russian prisoner in 1914. He explained the statement of his theorem to another Austrian mathematician who was in Russian captivity as well.

From the time of this mysterious meeting of two mathematical prisoners, the theorem lived in mathematical folklore until 1921 when the first proof of the Helly Theorem was published by the Austrian mathematician J. Radon (with reference that the theorem belonged to Helly). And in 1923 E. Helly published his own proof (different from Radon's!). Today, dozens of proofs of the Helly Theorem are known. Numerous papers have been written in which the theorem is applied.

Let us start the account of the subject. In Figure 18.1, four figures are drawn, such that every three of them have a common point. But there is no common point of all four figures.

Note, however, that one of the figures is not convex. And this is essential: if every three of four *convex* figures have a common point, then there exists a point that belongs to all four figures.

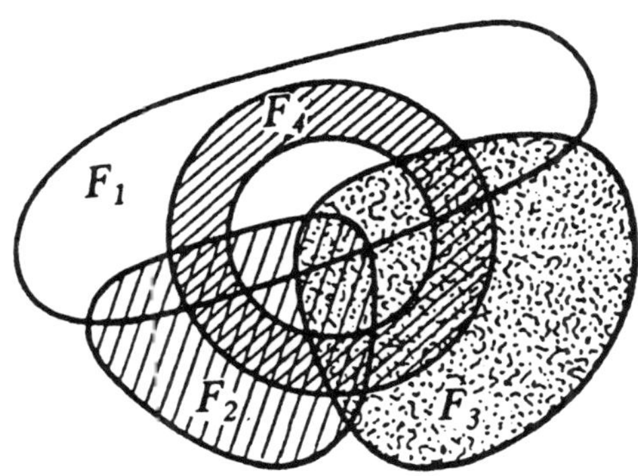

Figure 18.1

Exercise 18.1. Four convex figures are given in the plane. Prove that if every three of them have a non-empty intersection, then the intersection of all four figures is non-empty as well (the definition of intersection is found in Section 8, p. 76).

Exercise 18.2 (*The Helly Theorem for the Plane*). A finite family F_1, ..., F_s of convex figures is given in the plane. Prove that if every three of them have a non-empty intersection, then the intersection $F_1 \cap ... \cap F_s$ of all these figures is non-empty as well.

Note that in the statement of Exercise 18.2 we assume that every *three* of the given convex figures have a non-empty intersection. But 3 is equal to $n + 1$, where $n = 2$ is the *dimension* of the plane. This prompts the statement of the Helly theorem in space (that has dimension $n = 3$): *if every four of the given convex figures F_1, ..., F_s in the space have a common point, then the intersection of all the figures is non-empty.*

And here is the statement of the theorem for n-dimensional space R^n.

The Helly Theorem. *Let F_1, ..., F_s be convex figures in R^n. If each $n + 1$ of these figures have a common point, then the intersection $F_1 \cap ... \cap F_s$ is non-empty.*

Certainly, we suppose in this statement that s is a large number (at any rate, $s \geq n + 1$). The statements for the plane (Exercise 18.2) and for three-dimensional space are particular cases of this general theorem.

What is the "profit" of the Helly Theorem? From numerous examples of its application we will list several in the following exercises.

Exercise 18.3. There are s points in the plane (s is a positive integer), such that every three of them are contained in a disk of radius d. Prove that all the points are contained in a disk of radius d.

Exercise 18.4 (*Theorem of H. Jung*). Let M be a convex polygon in the plane, such that the distance between every two of its vertices does not exceed d. Prove that M if contained in a disk of diameter $d\dfrac{2}{\sqrt{3}}$.

In the above statements we considered only *finite* families of convex figures. It is interesting to find out whether the Helly Theorem holds for an *infinite* family of convex figures. The answer is "yes", if all the figures are closed and one of them is *compact* (that is, closed and bounded).

Exercise 18.5. Let F_1, ..., F_s... be a sequence of convex figures in the plane, each of which is closed and at least one of which is compact. Prove that if every three of the figures F_1, ..., F_s,... have a non-empty intersection, then the intersection $F_1 \cap ... \cap F_s...$ of all the figures is non-empty as well.

Exercise 18.6. Is the condition "each of the figures is closed" in Exercise 18.5 essential?

Exercise 18.7. Is the condition "one of the figures is compact" in Exercise 18.5 essential?

In Exercise 18.5, we had an infinite sequence of convex figures. But the assertion of this exercise holds for every family of convex figures, not necessary enumerated by positive integers 1, 2, ...:

> *If we have an infinite family of closed convex figures in the plane, one of which is compact, and if each three of them have a common point, then the intersection of all figures of the family is non-empty.*

This is the **Helly Theorem for an infinite family of convex figures.** The statement for the space R^3 is similar.

There is a beautiful conjecture associated with the above theorem. The great Paul Erdös suggested that we include it here for your (and our) enjoyment. Thank you, Paul!

Conjecture 18.1. *Given an infinite family of closed convex figures in the plane, one of which is compact. If among any four figures there are three figures with a point in common, then there is a finite set S (consisting of N points) such that every given figure contains at least one point from S.*

Moreover, the positive integer N is an absolute constant, i.e., it is one and the same for all famlies of figures that satisfy the above conditions.

To make it more challenging, we are offering \$25 for the first proof of Conjecture 18.1 or a counter example disproving it.

In view of the Helly Theorem for an infinite family of convex figures, the statements of Exercises 18.3 and 18.4 can be generalized.

Exercise 18.3'. Let M be a figure in the plane, such that every three of its points are contained in a disk of radius d. Then the whole figure M is contained in a disk of radius d.

Exercise 18.4'. Each figure of diameter d in the plane is contained in a disk of diameter $\dfrac{2}{\sqrt{3}}d$

Exercise 18.8. A bounded figure M of area S is given in the plane. Prove that there exists a point q in the plane, such that each line through q divides M into two parts, each of which has area not less than $\dfrac{1}{3}S$ (Figure 18.2).

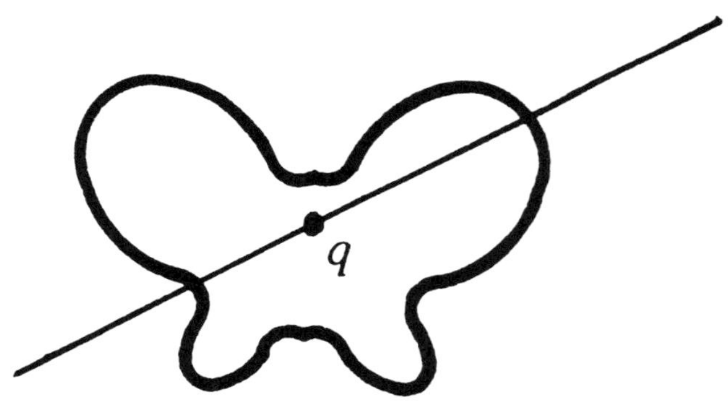

Figure 18.2

Exercise 18.9. A bounded convex figure M is given. Prove that there exists an interior point q of M such that every chord $[ab]$ of M through q is divided by q into two parts, each of which has a length not less than $\dfrac{1}{3}\,|\,ab\,|$ (Figure 18.3).

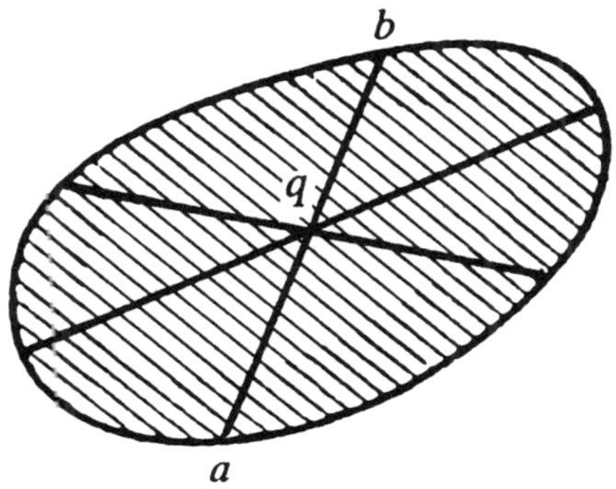

Figure 18.3

In the following exercises we will consider **_translates_** of a convex figure F, that is, figures obtained from F by translations (Figure 18.4).

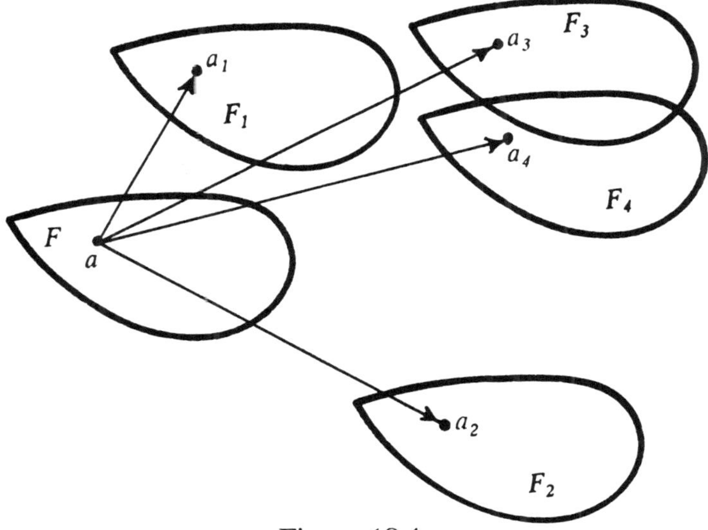

Figure 18.4

Exercise 18.10. Let $F_1, ..., F_s$ be translates of a parallelogram. Prove that if every two of the parallelograms $F_1, ..., F_s$ have a common point, then the intersection of all the parallelograms is non-empty.

What is the difference between the Helly Theorem and the previous exercise? In the Helly theorem we assume that every *three* of convex figures $F_1, ..., F_s$ have a common point and conclude that the intersection of all the figures is non-empty. In Exercise 18.10 we have a special case when $F_1, ..., F_s$ are translates of a parallelogram. In this case, it is sufficient to assume that every *two* of the figures $F_1, ..., F_s$ have a common point, and we can conclude that the intersection of all the figures is non-empty.

Is there a convex figure F distinct from a parallelogram with the same property: for any of its translates $F_1, ..., F_s$ if every two of them have a common point then the intersection $F_1 \cap ... \cap F_s$ is non-empty? The answer is "no". Confirming this fact is the next task for the reader:

Exercise 18.11. Prove that if F is a convex figure in the plane distinct from a parallelogram, then there exist three translates F_1, F_2, F_3 of F that pairwise have common points, but the intersection $F_1 \cap F_2 \cap F_3$ is empty.

The statement of the previous exercise is a particular plane case of the theorem that was established by the well-known Hungarian mathematician Szökefalvi-Nagy in 1954. Let us formulate this result.

The Szökefalvi-Nagy Theorem. *Let F be a convex figure in R^n with the following property: for any of its translates $F_1, ..., F_s$ if every two of them have a common point then the intersection $F_1 \cap ... \cap F_s$ is non-empty. Then F is a parallelotope (i.e., the n-dimensional analog of a parallelogram in R^2 and a parallelepiped in R^3).*

This theorem raises an interesting problem:

The Szökefalvi-Nagy Problem. *Let r and n be positive integers and $r \leq n$. Describe all convex bodies F in R^n with the following property: for any translates $F_1, ..., F_s$ of F, if every $r + 1$ of them have a common point, then the intersection $F_1 \cap ... \cap F_s$ is non-empty.*

We will denote this property by *Sz(r)*.

In the case $r = n$ the solution is given by the Helly Theorem: for $r = n$ the property $Sz(n)$ holds for every convex body in R^n. In the case $r = 1$ the solution is given by the Szökefalvi-Nagy Theorem: for $r = 1$ the property $Sz(1)$ holds if and only if F is a parallelotope in R^n. So, it would be interesting to consider this problem for the intermediate cases: $1 < r < n$.

The Szökefalvi-Nagy problem is not completely unsolved. Some partial results have been established. For centrally symmetric convex bodies a solution was obtained by V. Boltyanski in 1976 [B3] in the case $n = 2$, and by the Hungarian mathematician J. Kincses in 1987 [Ki] in the cases $n = 3$ and $n = 4$. Without the assumption of the central symmetry, a solution was recently obtained by V. Boltyanski in [B6]. This means that in the three-dimensional space R^3 we have a complete solution (that is, for $r = 1$, 2 or 3). But for $n > 3$ a complete solution is still unknown.

In conclusion, we would like to describe some interesting facts in R^3 obtained in [B3] and [B6].

A centrally symmetric body F in R^3 possesses the property $Sz(2)$ if and only if it is *a cylinder*, that is, the vector sum of a two-dimensional convex figure and a segment (Figure 18.5). In other

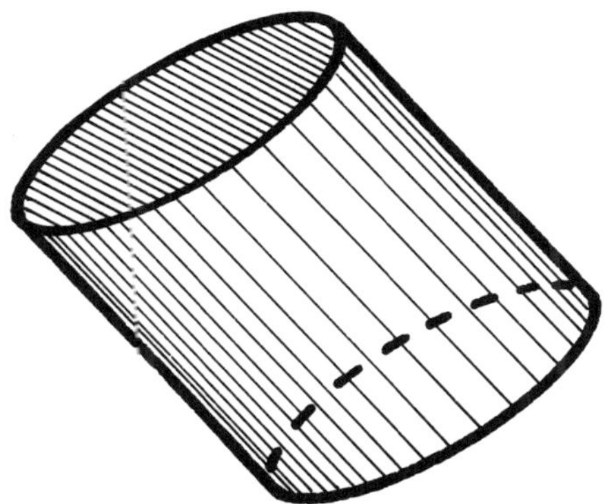

Figure 18.5

words, let F be a centrally symmetric convex body in R^3 with the following property: for any translates F_1, ..., F_s of F, if every three of them have a common point, then the intersection $F_1 \cap ... \cap F_s$ is non-empty. Then F is a cylinder with a centrally symmetric base.

214

Another interesting result is related to pyramids with quadrangular base (Figure 18.6). If F_0 is a pyramid with a parallelogram as its base, then F_0 possesses the property $Sz(2)$. In other words, for any translates F_1, ..., F_s of the pyramid, if every three of them have a common point, then the intersection $F_1 \cap ... \cap F_s$ is non-empty.

But if the base of a pyramid F is a quadrangle distinct from a parallelogram, then this assertion is not true! This means that there are four translates F_1, F_2, F_3, F_4 of F, such that every three of them have a common point, but the intersection $F_1 \cap F_2 \cap F_3 \cap F_4$ is empty (Can you prove it?).

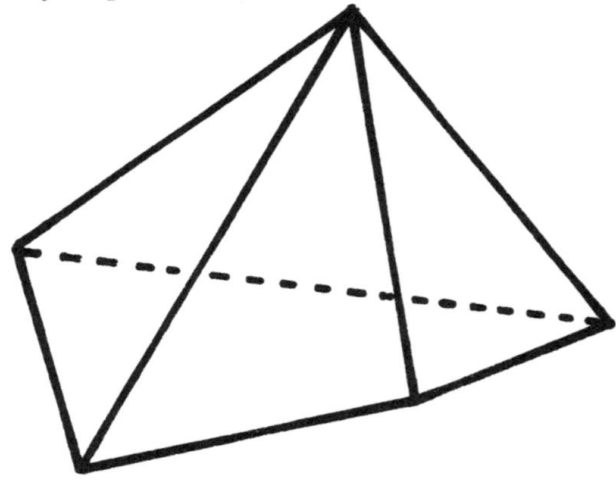

Figure 18.6

In Figures 18.7 and 18.8 some other convex bodies are shown that have the property $Sz(2)$.

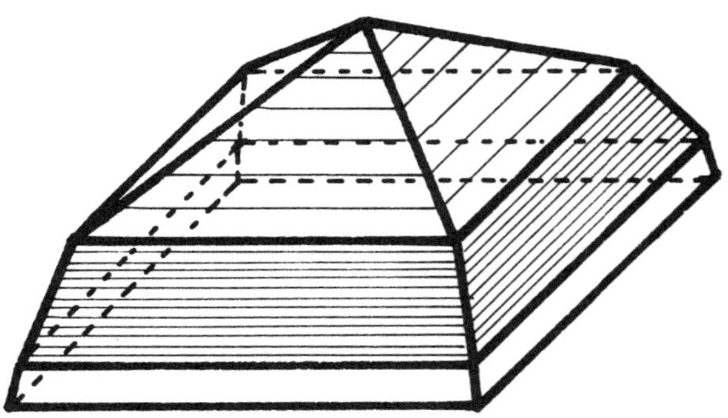

Figure 18.7

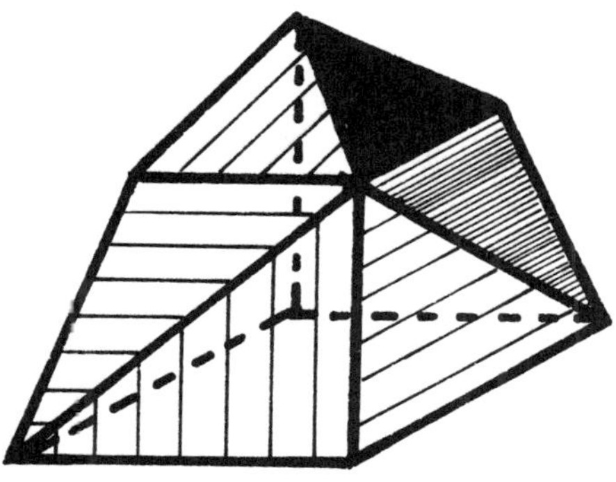

Figure 18.8

Solutions of Exercises

18.1. Let a_j be a point that belongs to all the convex figures F_1, F_2, F_3, F_4 except possibly F_j (that is, a_1 belongs to F_2, F_3, *and* F_4, and so on). We consider two cases:

i) One of the points a_1, a_2, a_3, or a_4 is contained in the triangle with the vertices in one of the other three points, say, a_1 is contained in the triangle $a_2\,a_3\,a_4$ (Figure 18.9).

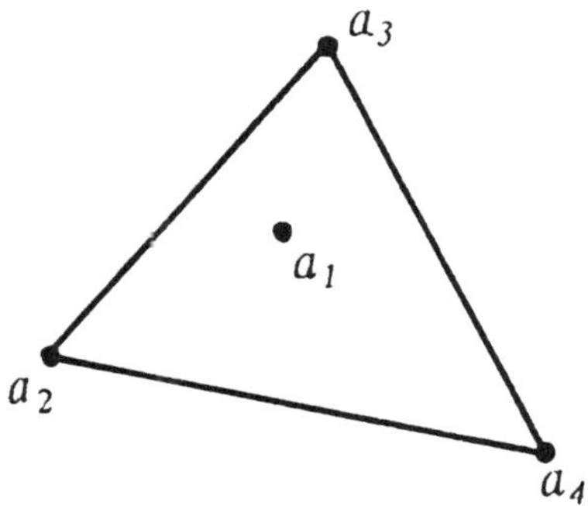

Figure 18.9

Each of the points a_2, a_3, a_4 belongs to the figure F_1. Consequently, the triangle $a_2 a_3 a_4$ is contained in F_1. Hence, the point a_1 is contained in F_1 too. Moreover, a_1 belongs to each of the figures F_2, F_3, F_4. Thus a_1 is a common point of all the figures F_1, F_2, F_3, F_4.

ii)   None of the points a_1, a_2, a_3, a_4 is contained in the triangle with the vertices in other three points, that is a_1, a_2, a_3, and a_4 are the vertices of a convex quadrangle (Figure 18.10).

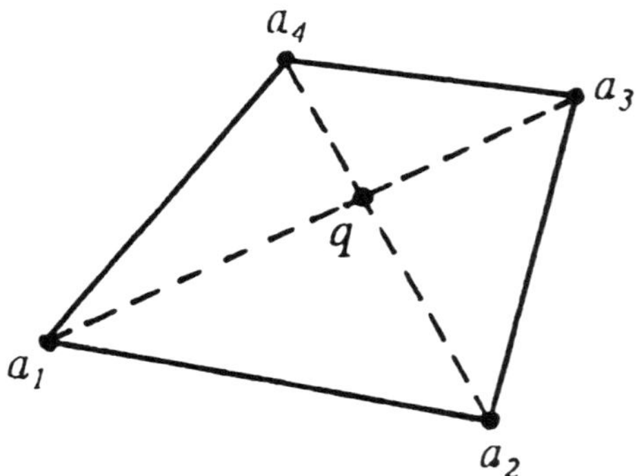

Figure 18.10

Without loss of generality we can assume that a_1 and a_3 are the opposite vertices of the quadrangle. The point a_1 belongs to each of the figures F_2 and F_4. The point a_3 belongs to each of the figures F_2 and F_4 too. Consequently, the segment $[a_1,a_3]$ is contained in each of the figures F_2 and F_4. Similarly, the segment $[a_2,a_4]$ is contained in each of the figures F_1 and F_3. Thus, the intersection point of the diagonals $[a_1,a_3]$ and $[a_2,a_4]$ belongs to each of the figures F_1, F_2, F_3, and F_4.

The cases i) and ii) address all possible situations when the points a_1, a_2, a_3, a_4 are not all on a line. If they are all on a line, then the proof is not difficult. (Try it on your own!)

18.2.   Let us proceed by induction on s. If $s = 4$, then the assertion is true (Exercise 18.1). Assume that for some $s \geq 4$ the assertion is true. For $s + 1$ convex figures F_1, ..., F_s, F_{s+1}, such that every three of them have a non-empty intersection, we denote the intersection $F_j \cap F_{s+1}$ by G_j (where $j = 1, ... s$). Then every three of the convex figures G_1, ..., G_s have a common point (indeed, $G_i \cap G_j \cap G_k$ is equal to $F_i \cap F_j \cap F_k \cap F_{s+1}$, and this intersection is non-empty, by virtue of the result of Exercise 18.1). Hence, by the inductive assumption, the intersection $G_1 \cap ... \cap G_s$ is non-empty. But this intersection coincides with $F_1 \cap ... \cap F_s \cap F_{s+1}$.

18.3. We have to prove that there exists a point q (the center of the desired disk), such that the distance from each given point to q does not exceed d. In other words, we have to find a point q that belongs to each disk of radius d with the center at one of the given points.

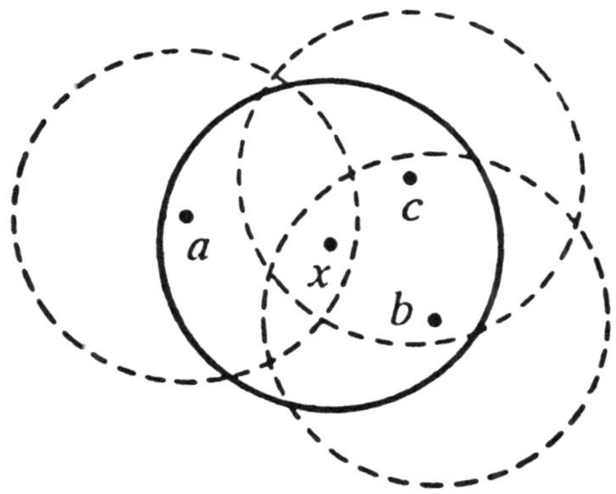

Figure 18.11

According to the Helly Theorem, it is sufficient to prove that every *three* of these disks have a common point. But we know that every three given points a,b,c are contained in a disk of radius d (Figure 18.11). The center x of this disk is a required common point of the disks of radius d with the centers in a, b, and c.

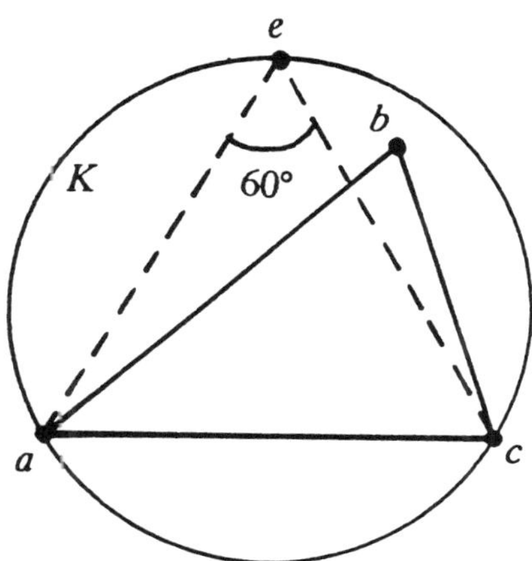

Figure 18.12

18.4. By the result of Exercise 18.3, it is sufficient to prove that every three vertices of the polygon M are contained in a disk of diameter $d\frac{2}{\sqrt{3}}$. Let a, b, and c be three vertices of M, and $[a,c]$ be the largest side of the triangle abc. The angle abc is not smaller than $60°$. So, the point b belongs to the disk K circumscribed about the equilateral triangle with the side $[a,c]$ (Figure 18.12). But the diameter of this disk is equal to $|ac|\frac{2}{\sqrt{3}}$, that is, it does not exceed $d\frac{2}{\sqrt{3}}$.

18.5. Without loss of generality, we can assume that the figure F_1 is compact. According to the Helly Theorem, the intersection of the figures $F_1, ..., F_m$ is a non-empty convex figure. Moreover, it is compact. We denote this intersection by G_m. Then $G_1, G_2, ...$ is a decreasing sequence of non-empty compact figures. By virtue of Example 8.4 there exists a point that belongs to each figure G_m, and consequently, to each figure F_m.

18.6. Let F_m be an open disk (that is, a disk without its boundary points) of radius 1 with the center $a_m = (\frac{1}{m},0)$ where $m = 1,2, ...$ Let, further, F_0 be a closed disk of radius 1 with the center $a_o = (2,0)$. We obtain a sequence $F_0, F_1, ..., F_m, ...$ of convex figures in the plane where F_0 is compact (Figure 18.13). Any finite number of the figures $F_0, F_1, ..., F_m, ...$ has a non-empty intersection, but the intersection of *all* these figures is empty.

Note that the intersection of the corresponding *closed* disks consists of only the point $(1,0)$.

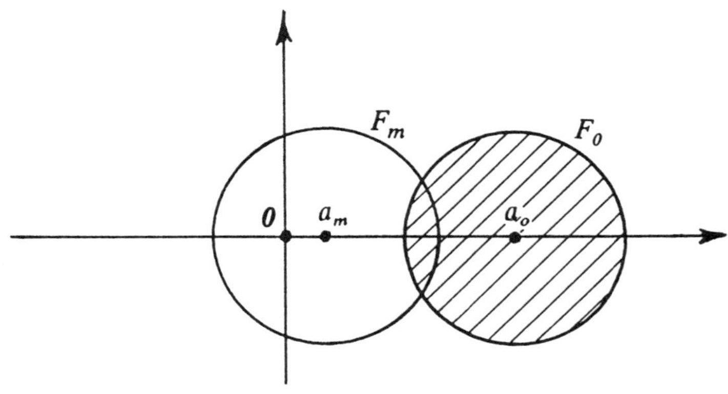

Figure 18.13

18.7. Let us consider the family of all "right" closed halfplanes defined by vertical lines (Figure 18.14). None of these figures is compact. And the intersection of all figures of this infinite family is empty (since there is in the family a halfplane moved as far away to the right as we like).

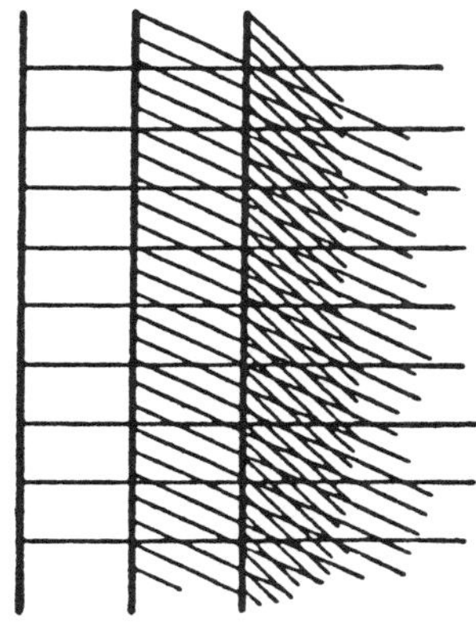

Figure 18.14

18.8. Let us consider all the closed halfplanes, each of which contains a part of M with the area exceeding $\frac{2}{3}S$. So the part of M situated outside of such a halfplane has an area *smaller* than $\frac{1}{3}S$. This means that every three of the considered halfplanes have a non-empty intersection (Can you see why?). Now let us take a disk K containing M and replace each halfplane H of the considered family by the intersection $H \cap K$ (Figure 18.15). We obtain

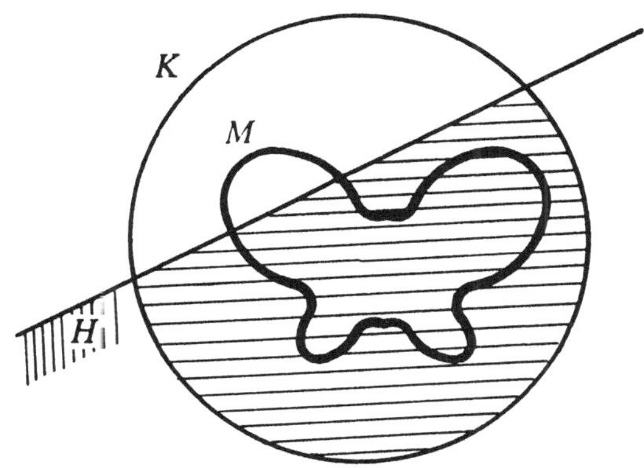

Figure 18.15

an infinite family of compact, convex figures in the plane, each three of which have a common point. Consequently, there exists a point q that belongs to each figure of the family. This point q is a required one. Indeed, let L be a line through q (Figure 18.16). If a half-plane with the boundary L were to contain a part of M with the area *smaller* than $\frac{1}{3}$S, then a parallel line L' would exist, such that a halfplane with the boundary L' would contain a part of M with area greater than $\frac{2}{3}$S and would not contain q, contradicting the construction of q. This completes the solution.

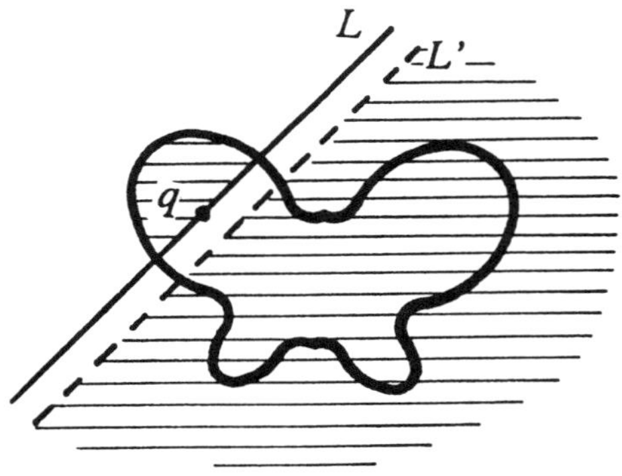

Figure 18.16

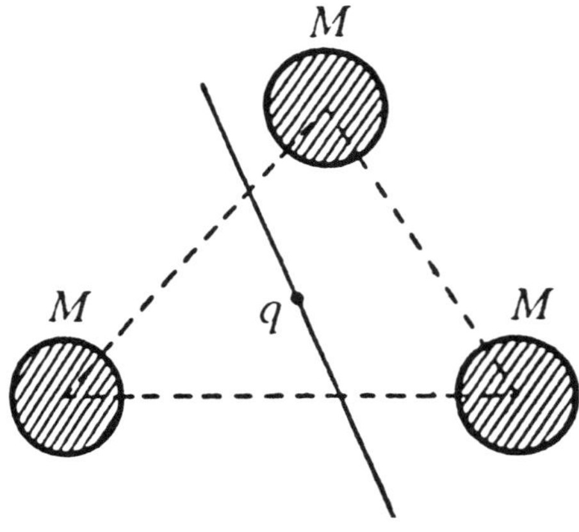

Figure 18.17

Please note that the number $\frac{1}{3}$ in the statement of Exercise 18.8 cannot be replaced by a greater number. For example, if the figure M consists of three separate congruent disks (Figure 18.17), then the number $\frac{1}{3}$ can not be improved. On the other hand, if M is convex, then the number $\frac{1}{3}$ can be replaced by $\frac{4}{9}$.

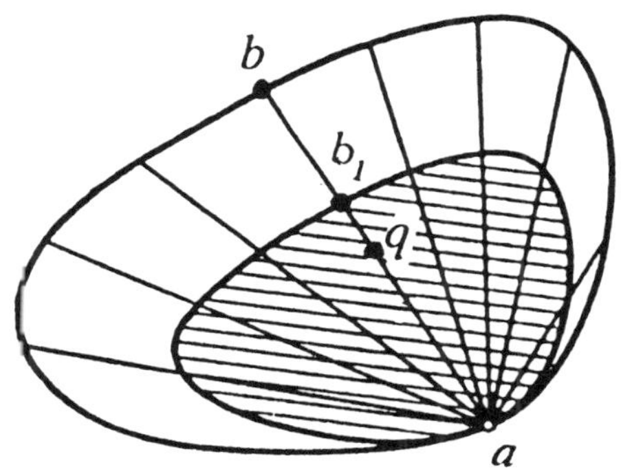

Figure 18.18

18.9. For every point a of M we define M_a as the figure homothetic to M with center a and ratio $\frac{2}{3}$ (Figure 18.18).

We consider the family of all the figures M_a where a belongs to M. If M_a, M_b, and M_c are three figures of this family (Figure 18.19), then the triangle abc is contained in M.

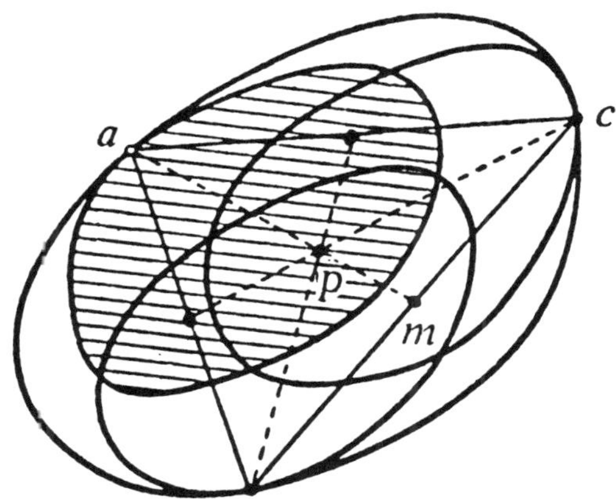

Figure 18.19

Then the median $[a,m]$ is contained in M and, consequently, the centroid p of the triangle abc (i.e., the point of intersection of medians) is contained in M_a. Similarly, p belongs to M_b and M_c. Thus, every three figures of the considered family have a common point. Moreover, each figure M_a is convex and compact. Hence, there exists a point q that belongs to every figure M_a of our family.

Let now $[a,b]$ be a chord through q (Figure 18.18), and b_1 be the image of b under the homothety with the center a and radio $\frac{2}{3}$. Then $[a,b_1]$ is a chord of M_a, and the point q belongs to the segment $[a,b_1]$ (since q is contained in M_a). Consequently,

$$|bq| \geq |bb_1| = \tfrac{1}{3}|ab|$$

Similarly, $|aq| \geq \tfrac{1}{3}|ab|$.

18.10. It is sufficient to prove that every *three* of the parallelograms considered have a non-empty intersection (then by virtue of the Helly Theorem, we can conclude that the intersection $F_1 \cap ... \cap F_s$ is non-empty as well).

So, let us prove that, for example, the intersection $F_1 \cap F_2 \cap F_3$ is non-empty. Assume the opposite, that is, the intersection $F_1 \cap F_2$ has no common points with the parallelogram F_3. Then there exists a line L parallel to one side of the parallelograms that separates $F_1 \cap F_2$ and F_3 (Figure 18.20).

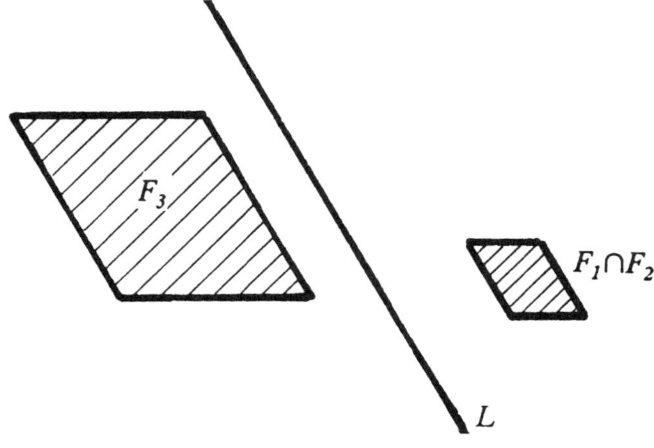

Figure 18.20

Let us replace each parallelogram F_i by the corresponding strip S_i with sides parallel to L. Then, as before, the line L separates $S_1 \cap S_2$ and S_3 (Figure 18.21).

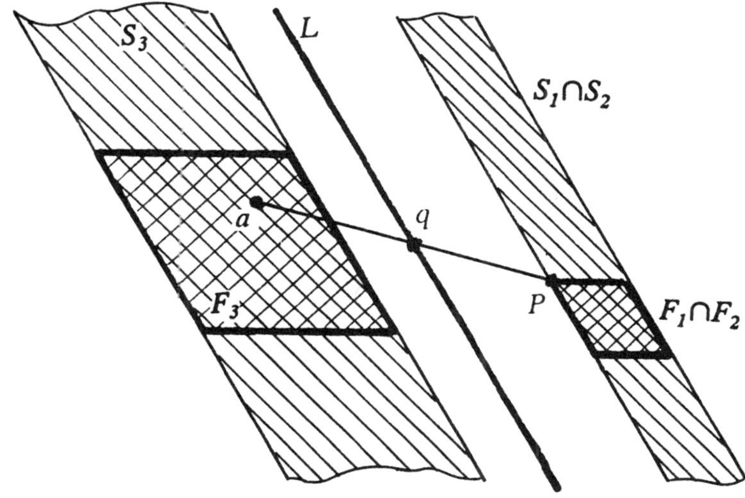

Figure 18.21

But we know that F_1 and F_3 have a common point a. Further, if we take a point p of $F_1 \cap F_2$, then both the points a and p belong to F_1. Consequently, the segment $[a,p]$ is contained in F_1. This segment has a common point q with the line L (since a and p are situated on the different sides of L). Thus, L has the common point q with F_1. Therefore, L is contained in S_1. Similarly, L is contained in S_2. Hence L is contained in the intersection $S_1 \cap S_2$, contradicting the construction of L. This contradiction shows that the intersection $F_1 \cap F_2 \cap F_3$ is non-empty.

18.11. Since F is distinct from a parallelogram, then there are three directions that illuminate the whole boundary of F (see Theorem 17.1). Six support lines of F parallel to these illuminating directions form a hexagon circumscribed about F (Figure 18.22). The points a, b, and c in Figure 18.22 *cannot belong* to F because otherwise they would not be illuminated.

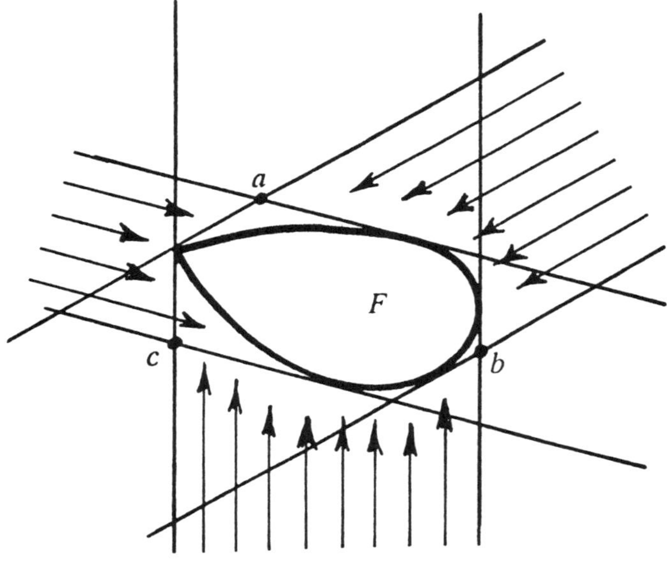

Figure 18.22

Let us now choose a point q and consider three translations through the vectors $\overrightarrow{aq}$, $\overrightarrow{bq}$ and $\overrightarrow{cq}$ (Figure 18.23).

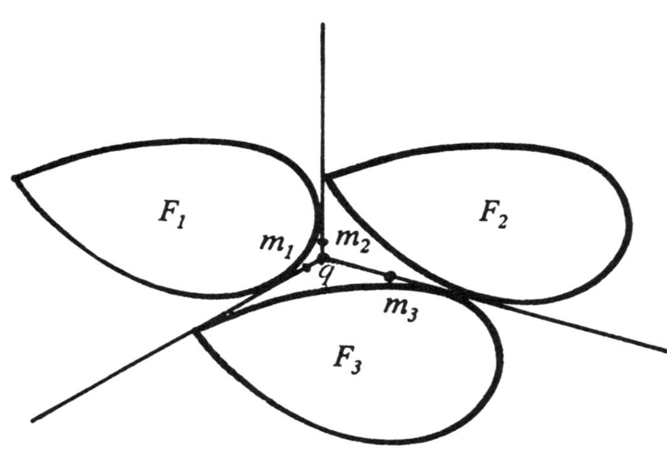

Figure 18.23

We obtain three translates F_1, F_2, and F_3 of F that don't contain the point q and have pairwise a common supporting ray emanating from q. Finally, we choose three points m_1, m_2, and m_3 on these rays closer to q than the figures F_1, F_2, F_3 and translate the figures F_1, F_2, F_3 to new positions such that the translated figures each contain two of the points m_1, m_2, or m_3, but do not contain q (can you see why it is always possible?) These new translates of F have pairwise common points m_1, m_2, or m_3, but the intersection of all three figures is empty.

BIBLIOGRAPHY

[B1]   V.G. Boltyanski. A problem about the illumination of the boundary of a convex body, *Izv. Moldavsk. Filiala Akad. Nauk SSSR*, 10(76), 1960, 79-86 (Russian).

[B2]   V.G. Boltyanski. O razbienii ploskikh figur na chasti men'shevo diametra (On partitioning plane figures into parts of smaller diameters). *Colloquium Math.* Warszawa, 21(2), 1970, 253-263 (Russian).

[B3]   V.G. Boltyanski. Generalization of a certain theorem of Szökefalvi-Nagy, *Dokl. Adad. Nauk SSSR*, 228(2), 1976, 265-268 (Russian).

[B4]   V.G. Boltyanski. Several theorems of combinatorial geometry. *Mat. Zametki*, 21(1), 1977, 117-124 (Russian).

[B5]   V.G. Boltyanski & P.S. Soltan. Combinatorial geometry and convexity classes. *Uspehi Mat. Nauk*, 33, 1(199), 1978, 3-42, 262 (Russian).

[B6]   V.G. Boltyanski. A new step in the solution of the Szökefalvi-Nagy problem (to appear).

[Bo1]   K. Borsuk. Über die Zerlegung einer Euklidischen n-dimensionalen Vollkugel in n Mengen. *Verh. Internat. Math. Kongr.* Zurich, No. 2, 1932, 192 (German).

[Bo2]   K. Borsuk. Drei Sätze über die n-dimensionale Sphäre. *Fund. Math.*, 20, 1933, 177-190 (German).

[BG]   V.G. Boltyanski & I. Ts. Gohberg. Results and Problems in Combinatorial Geometry, Cambridge University Press, Cambridge, 1985.

[DGK]   L. Danzer, B. Grünbaum, & V. Klee. Helly's theorem and its relatives. *Proc. Symp. Pure Math.*, 7 (Convexity), 101-180.

[E1]   H.G. Eggleston. Covering a three-dimensional set with sets of smaller diameter. *J. London Math. Soc.*, 30, v. 1, 1955, 11-24.

[E2]   H.G. Eggleston. *Convexity.* Cambridge University Press. London, 1958.

[F]   I. Fáry. On straight line representations of planar graphs. *Acta Sci. Math.*, *11*, 1948, 229-233.

[Ga]   D. Gale. On inscribing n-dimensional sets in a regular n-simplex. *Proc. Amer. Math. Soc.*, 4, 1953, 222-225.

[Go1]   S.W. Golomb. Checkerboards and Polyominoes, *Amer. Math. Monthly*, *61*, No. 10, 1954, 672-682.

[Go2]   S.W. Golomb. *Polyominoes.* Charles Scribner and Sons, New York, 1965.

[GK]   S.W. Golomb, D.A. Klarner. Covering a rectangle with L-tetrominoes, *Amer. Math Monthly*, *70*, v. 7, 1963, 760-761.

[G1]   B. Grünbaum. A simple proof of Borsuk's conjecture in three dimensions. *Proc. Cambridge Philos. Soc.*, 53, 1957, 776-778.

[G2]   B. Grünbaum. *Convex Polytopes (Pure and Appl. Math., v. 16).* John Wiley & Sons, New York, 1967.

[GS]   B. Grünbaum & G.C. Shephard. *Tilings and Patterns.* W.H. Freeman and Co., New York, 1987.

[Gu]   R.K. Guy. Crossing number of graphs. *Graph Theory and Applications.* Springer-Verlag, New York, 1972, 111-124.

[H1]   H. Hadwiger. Überdeckung einer Menge durch Mengen kleineren Durchsmessers. *Comm. Math. Helv.*, 18, 1945-46, 73-75 (German).

[H2]   H. Hadwiger. Überdeckung einer Menge durch Mengen kleineren Durchsmessers. *Comm. Math. Helv.*, 19, 1946-47, 72-73 (German).

[H3]   H. Hadwiger. *Altes und neues über konvexe Körper.* Birkhäuser, Basel-Stuttgart, 1955 (German).

[H4]   H. Hadwiger. Ungelöste Probleme, *Elem. der Math.*, 1957, 12, No. 20, 121 (German).

[He1]   E. Helly. Über Mengen konvexer Körper mit gemeinschaftlichen Punkten, *Jber. Deutsch. Math. Verein.*, 32 (1923), 175-176 (German).

[He2]   E. Helly. Über Systeme von abgeschlossenen Mengen mit gemeinschaftlichen Punkten. Monatsh. Math., 37, 1930, 281-302 (German).

[Hep]   A. Heppes. Térbeli ponthalmazok felosztása kisseb atméröjü részhalmazok összegére. Magyar tud akad. *Mat. es fiz tud közl.*, 7, 1957, 413-416 (Hungarian).

[HRW]   F. Harary, R.W. Robinson, & N. Warmald. Isomorphic Factorizations I: Complete Graphs. *Trans. Amer. Math. Soc.*, 242, 1978, 243-260.

[K]   A.B. Kharazishvili. K zadache osvetshenia. (On the problem of illumination). Soob AN *GSSR* 1973, 289-291 (Russian).

[Ki]   J. Kincses. The classification of 3- and 4-Helly dimensional convex bodies. *Geometriae Dedicata*, 22, 1987, 283-301.

[Kl]   D.A. Klarner. Packing a rectangle with congruent N-ominoes. *J. Combin. Theory*, 7, 1969, 107-115.

[Ko]   K. Kolodziejczuk. Borsuk's covering and planar sets with unique completion (to appear).

[Ku]   K. Kuratowski. Sur le probléme des courbes gauches en topologie. *Fund. Math.*, 16, 1930, 271-283 (French).

[LS]   L.A. Liusternik & L.G. Shnirelman. Topologicheskije Metody v Variazionnyh Zadachah (Topological Methods in Variational Problems). *M., Ohti.*, 1930 (Russian).

[P]   J.F. Pál. Ein Minimumproblem für Ovale. *Math. Ann.* 83, 1921, 311-319 (German).

[R]   J. Radon. Mengen konvexer Körper, die einen gemeinsamen Punkt enthalten. *Math. Ann.*, 83, 1921, 113-115 (German).

[S1]   A. Soifer. *Mathematics as Problem Solving.* Center for Excellence in Mathematical Education, Colorado Springs, CO 1987.

[S2]   A. Soifer. *How Does One Cut a Triangle?* Center for Excellence in Mathematical Education, Colorado Springs, CO, 1990.

[S3]   A. Soifer. Kletchatye doski i polyomino (Checker boards and polyominoes). *Kvant,* 11, 1972, 2-10 (Russian).

[S4]   A. Soifer. *Mathematics as Problem Solving II.* Center for Excellence in Mathematical Education, Colorado Springs (to appear).

[SL]   A. Soifer & E. Lozansky. Pigeons in every pigeonhole. *Quantum,* January, 1990, 25-26, 32.

[SS]   A. Soifer & S.G. Slobodnik, Problem M236, *Kvant,* No. 12, 1973, p. 29 (Russian).

[SN]   B. Szökefalvi-Nagy. Ein Satz über Parallelverschiebungen konvexer Körper. *Acta Sci. Math.,* 15, 1954, 169-177 (German).

[T]   S.L. Tabachnikov. Considerations of continuity. *Quantum,* May 1990, 8-12.

[W]   K. Wagner. Bemerkungen zum Vierfarben problem, *Jber. Deutsch. Math. Verein.,* 46, 1936, 21-22 (German).

[Wa]   D.W. Walkup. Covering a rectangle with T-tetrominoes, *Amer. Math. Monthly,* 72, v. 9, 1965, 986-988.

[Y]   K. Yoneyama. Theory of continuous set of points. *Tôhoku Math. J.,* 12, 1917, 43-158.

[YB]   I.M. Yaglom & V.G. Boltjanski. *Vypuklye Figury* (Convex Figures). GITTL, Moscow, 1951 (Russian).

English translation: *Convex Figures.* Holt, Rinehart and Winston, New York, 1961.

INDEX

angular boundary point 147

Barbier Theorem 168

bigraph 94

Blaschke Theorem 149

Bolzano-Weierstrass Theorem 71

Bolzano-Weierstrass Theorem for
   Compact Figures 79

Borsuk Conjecture 149

Borsuk Number 158

Borsuk Theorem 156

boundary (of a figure) 140

boundary point 138

bounded (sequence of figures) 79

bounded point set (in the plane) 72

bounded subset (of the real line R) 71

chain 129

closed figure 139

closed set 74

compact figure 140

compact set 65

complete bigraph 95

complete graph 95

column cyclic k-coloring 28

complement (of a graph) 94

congruence modulo n 27

container (of a compact figure) 79

converging sequence (of compact figures) 79

convex figure 136

convex hull 141

convex polygon 142

convex polyhedron 143

cross-cut (of a convex figure) 83

crossing number (of a graph) 121

cutting a graph 111

cycle 129

cycle of acquaintances 93

cylinder 50

diagonal cyclic k-coloring 28

diameter 155

diametral chord 167

direct product of diagonal and column coloring 58

disk 73

distance (between two compact figures) 78

domino 1

ε-closeness (of a point to a figure) 77

ε-extension (of a figure) 78

edge (of a graph) 93

elementary subdivision (of a graph) 119

Euler Formula for Graphs 118

Euler Formula for Polyhedra 119

exact lower bound 72

exact upper bound 72

232

experiment 7

exterior region (of a plane graph) 117

extremal point (of a compact convex figure) 143

figure of constant width 164

graph 93

Hadwiger Covering Problem 196

Hadwiger Problem 195

Helly Theorem 208

Helly Theorem for an Infinite Family of
    Convex Figures 209

Helly Theorem for the Plane 208

homeomorphic graphs 119

homothety 196

homothetic image 196

h-hull 180

illumination 180

illumination problem 180

incidence (of a vertex and an edge of a graph) 93

Infinite Pigeonhole Principle 71

interior (of a figure) 140

interior point 138

Intermediate Value Theorem 75

intersection (of figures) 76

intersection index (of two chains) 129

intersection index (of two segments) 128

invariant 6

invariants (of graphs) 102

inverse (translation) 54

isomorphic graphs 101

isoperimetric problem 82

Jordan Theorem 128

L-tromino 15

limit (of a sequence of compact figures) 79

limit point of a set (in the plane) 73

limit point (of a subset of the real line *R*) 71

linear *k*-omino 25

linear tromino 5

minimal rectangle (with respect to a
   polyomino) 39

Möbius band 51

monomino 1

n-fold rotational symmetry (of a figure) 37

n-fold rotational symmetry (of a tiling) 38

n-omino 3

non-diametral chord 186

order (of a graph) 105

orientation (of a polyomino) 40

packing (a parallelepiped) 57

parallel (step-lines) 54

periodic (tiling of an infinite strip) 54

periodic (tiling of the plane) 56

Pigeonhole Principle 65

planar graph 117

plane graph 117

Ramsey number 95

rectilinear crossing number (of a graph) 131

region (of a plane graph) 117

regular boundary point 147

Reuleaux Triangle 165

separable figures 149

skeleton (of a polyhedron 117

smaller copy 196

step-line 54

strictly separable figures 149

subgraph 94

support line 144

support plane 150

Szökefalvi-Nagy Theorem 212

Szökefalvi-Nagy Problem 212

tetromino 14

tiling 1

torus 50

translates 211

tromino 3

vector addition of convex figures 170

vertex (of a convex polygon) 143

vertex (of a graph) 93

Weierstrass Theorem 74

NOTATIONS

$a_1 \equiv a_2 \ (mod \ n)$ 27

$C \ (m,n)$ 50

$T \ (m,n)$ 50

$M \ (m,n)$ 51

$[a,b]$ 71

$M_1 \cap M_2 \cap ...$ 76

$d \ (M_1, M_2)$ 78

$\overline{G}$ 91/94

$deg \ v$ 94

Σ 94

$K_{n_1 n_2}$ 95

K_n 95

$r(m,n)$ 95

$G_1 = G_2$ 102

$|G|$ 105

$\dfrac{n}{2}$ 108

$G(P)$ 117

$v(G)$ 121

$\overline{v}(G)$ 121

$I(X,Y)$ 128

$int \ M$ 140

$bd \ M$ 140

$conv \ X$ 141

$|ab|$ 151

R^2 156

R^3 156

R^n 156

$a(F)$ 158

$F_1 + F_2$ 170

F^* 180

$c(F)$ 180

$b(F)$ 197

$Sz(r)$ 212

COMMENTS
NEW SOLUTIONS AND PROBLEMS
(Feel free to add pages as necessary)

Send to: Alexander Soifer
          Center for Excellence in Mathematical Education
          885 Red Mesa Drive
          Colorado Springs, CO 80906, USA

NOTES

NOTES

NOTES

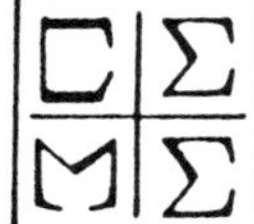

Center for Excellence in
Mathematical Education

885 Red Mesa Drive
Colorado Springs, CO 80906 USA
Phone: (719) 576-3020

ORDER FORM

SHIP TO:

Name: _______________________________

School: _______________________________

Address: _______________________________

Cty/St/Zip: _______________________________

Phone: _______________________________

BILL TO: (If different from "Ship to")

Name: _______________________________

School: _______________________________

Address: _______________________________

Cty/St/Zip: _______________________________

Phone: _______________________________

SHIP THE FOLLOWING:

QUAN	DESCRIPTION	PRICE PER ITEM	TOTAL AMOUNT
	Mathematics as Problem Solving	$15.95	
	How Does One Cut a Triangle?	$18.95	
	Geometric Etudes in Combinatorial Math	$24.95	
	Merchandise Total		
	Discount (see page ii)		
	Subtotal		
	Tax (if applicable)		
	Shipping & Handling (10% of order, $2.95 minimum. Foreign orders: 20%)		
	TOTAL		

METHOD OF PAYMENT

INDIVIDUALS AND INSTITUTIONS

☐ Check or money order enclosed. (Include shipping charges; Colo. residents add sales tax.)

SCHOOLS

☐ Please bill. (Attach signed P.O.) Purchase Order Number

☐ Check or money order enclosed.

Orders of 5 copies or more qualify for a 10% discount

* Orders must be prepaid. Unless net 30 days policy has been approved.
* All prices and terms are those in effect February 1, 1990 and are subject to change without notice.
* Orders are shipped best way. Our regular method of shippment is U.S. mail or United Parcel Service. If you require a special method of shippment(AIR MAIL for example), an additional cost will be incurred.

Center for Excellence in Mathematical Education

885 Red Mesa Drive
Colorado Springs, CO 80906 USA
Phone: (719) 576-3020

ORDER FORM

SHIP TO:

Name: _______________________________

School: _______________________________

Address: _______________________________

Cty/St/Zip: _______________________________

Phone: _______________________________

BILL TO: (If different from "Ship to")

Name: _______________________________

School: _______________________________

Address: _______________________________

Cty/St/Zip: _______________________________

Phone: _______________________________

SHIP THE FOLLOWING:

QUAN	DESCRIPTION	PRICE PER ITEM	TOTAL AMOUNT
	Mathematics as Problem Solving	$15.95	
	How Does One Cut a Triangle?	$18.95	
	Geometric Etudes in Combinatorial Math	$24.95	

Merchandise Total	
Discount (see page ii)	
Subtotal	
Tax (if applicable)	
Shipping & Handling (10% of order, $2.95 minimum. Foreign orders: 20%)	
TOTAL	

METHOD OF PAYMENT

INDIVIDUALS AND INSTITUTIONS

❏ Check or money order enclosed. (Include shipping charges; Colo. residents add sales tax.)

SCHOOLS

❏ Please bill. (Attach signed P.O.) Purchase Order Number

[   ]

❏ Check or money order enclosed.

Orders of 5 copies or more qualify for a 10% discount

* Orders must be prepaid. Unless net 30 days policy has been approved.
* All prices and terms are those in effect February 1, 1990 and are subject to change without notice.
* Orders are shipped best way. Our regular method of shippment is U.S. mail or United Parcel Service. If you require a special method of shippment(AIR MAIL for example), an additional cost will be incurred.